TRASH*formations*

PAINTED TREASURES FROM SALVAGED STUFF

Jennifer R. Ferguson and Judith A. Skinner

Martingale®
& COMPANY

Trashformations: Painted Treasures from Salvaged Stuff
© 2003 Jennifer R. Ferguson and Judith A. Skinner

Martingale & Company
20205 144th Avenue NE
Woodinville, WA 98072-8478
www.martingale-pub.com

CREDITS

President • Nancy J. Martin
CEO • Daniel J. Martin
Publisher • Jane Hamada
Editorial Director • Mary V. Green
Managing Editor • Tina Cook
Technical Editor • Candie Frankel
Copy Editor • Leslie Phillips
Design Director • Stan Green
Cover and Text Designer • Shelly Garrison
Illustrator • Lisa McKenney
Photographer • Brent Kane

Mission Statement
Dedicated to providing quality products
and service to inspire creativity.

Printed in China
08 07 06 05 04 03   8 7 6 5 4 3 2 1

Library of Congress Cataloging-in-Publication Data

Ferguson, Jennifer R.
  Trashformations : painted treasures from salvaged stuff / Jennifer R. Ferguson, Judith A. Skinner.
    p. cm.
    ISBN 1-56477-512-7
  1. Painting. 2. Decoration and ornament. 3. House furnishings. 4. Salvage (Waste, etc.) I. Skinner, Judith A. II. Title.
  TT385.F48 2003
  745.7'23—dc21
                                        2003013774

DEDICATION

From Jennifer

To my parents, Barbara and Leo.

Mom, I thank you for all that you have given me, much more than words could ever express. The love you give and the sacrifices you have made over the years—thank you! Dad, I wish you were here to share in the joy of the creation of this book, but I know you are still keeping an eye on me. Keep watching!

From Judy

To my mother, Esther.

I know my mom watches over me, but I can't help but wish she were here to share this with me. My mom gave me so much, I want this book to be for her. Thanks, Mom!

ACKNOWLEDGMENTS

We would like to express our immense gratitude to everyone who helped create this book:

- Martingale & Company, for publishing *Trashformations*
- Mary Green, Donna Lever, Terry Martin, and Shelley Santa, for their continued support and incredible help
- Jay and Mary Jones, for their wonderful woodworking talents and friendship
- Patsy Crow, for allowing us to use her wonderful backyard to photograph some of the projects
- DecoArt, for supplying us with wonderful products to create with
- Royal Brush Company, for supplying us with artist's brushes
- Scarab Glass Works, Fresno Ag Hardware, and Imperial Glass Company, for all their help with the glass projects in this book
- Stan Green and Brent Kane, for their wonderful photography. We couldn't ask for a better team—they both make our projects look incredible!

And finally, we thank our families: Jennifer's husband, Jim, and their children, Ashley and Tyler; and Judy's husband, Don, and their children and their families—Donnie, Joshua, Presli, Rex, Joanie, Corri, Bob, and Samantha. We thank you all for your continued support and endless love. We wouldn't be able to do all this without you.

Contents

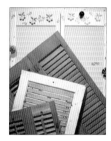

Introduction

Do you have an old door or window that you've been saving—but you're not sure what for? Has a kitchen or bathroom remodeling left you wondering what to do with those old drawers and cabinet doors? Do you have a passion for salvage yards and yard sales? Or maybe you're just a "junker" at heart?

If so, *Trashformations* is the book for you. We've put our creative thinking and love for salvage to work, coming up with 20 step-by-step projects to help you transform your finds into new, functional pieces. A little imagination, common household tools, and some paint are all it takes to transform an old screen door and a couple of newel posts into a potting bench. Or how about turning a trio of paneled doors into an elegant personal dressing screen? You've heard of upside-down cake, but have you ever seen an upside-down table, made by inverting a drawer on a pedestal base?

Whether you are a novice painter or an experienced artist, you can create every one of these wonderful projects with ease.

Begin your trashformation journey by browsing through our opening chapters. You'll learn how to find items to salvage, what tools and supplies are needed, and how to prepare different surfaces for painting. There's even a section on cutting and installing replacement glass. Next, you'll explore the world of decorative painting techniques. Our step-by-step project instructions list the specific paint colors and stencils that we used. Of course, you may want to combine painting techniques, designs, and colors from several projects to create your own unique pieces. That's the art of giving new life to salvaged materials.

Whether or not you officially qualify as a "junker," these incredible creations are sure to spark your creativity. We love giving old castoffs new life. Come join the revival!

Hunting for Salvage

Finding items to salvage is part work, part luck, and always fun. Start by picking up your local telephone directory and looking under "Salvage" to see if there are any nearby salvage yards you can visit. A salvage yard will have many things for you to pick through: old windows, doors, shutters, and cabinets, to name a few. You might also check on the internet for salvage yards that are close enough for a day trip. Most of our junking trips over the past years have been planned around road trips to conventions, guest appearances, and teaching engagements. We live for our junking trips!

Flea markets and yard sales are also possibilities, but the "inventory" is less predictable. Yard sales and charity rummage sales often yield the best prices, especially toward the end of the sale when unloading the stuff becomes a priority. On the other hand, if you don't get there early, you may miss out on a really good item.

Some of our best finds have been free for the asking. Remodeling projects and demolition sites, especially in older neighborhoods, are good places to inquire. Tell your friends and family what you are looking for, and you'll multiply your leads. If we spot something out on the sidewalk on trash pickup day, we're not shy about snapping it up before the collection truck arrives. As a general rule, we never leave home without a hammer, screwdriver, chisel, and work gloves in the car. This way, if we drive by a house ready for stripping or spot a good junk pile on the side of the road, we're prepared.

For an all-day junking excursion, we recommend planning ahead. If you're shopping for big items like doors and windows, leave the Corvette at home. You need a vehicle large enough to transport your prizes back to home base. Other road trip essentials include coffee (Starbucks gets us going!), bottled water, and snacks. Who wants to waste valuable junking hours tracking down a place to eat?

Always pack plenty of cash. Some places may not take credit cards or checks, and the nearest ATM might be miles away. Besides, your best price will always be with a cash transaction. Haggling over the price is half the fun. We always like to think we got the better end of the bargain—whether we actually did or not. Even in an antique shop, where the pricing is apt to be higher, it never hurts to ask, "Is this your friendliest price?"

Don't be afraid to really scavenge and get dirty as you hunt for pieces you can use. Personally, we think if you don't get a little dirty when you're junking, you're not getting the best deal. Dress comfortably and casually. Junking is different from shopping at Macy's. Overdressing can send the wrong message and hurt your bargaining power.

We hope you enjoy reading about some of our scavenging successes in the project section of the book. One thing we can guarantee: Once the scavenging bug hits you, you will never look at a discarded door or window in the same way again.

The old windows and doors shown here are in their "found" condition.

No Time to Salvage

If picking your way through a junk pile is not your thing, there's no salvage yard nearby, or you're simply pressed for time, here's another option. A home improvement store offers "one-stop shopping" for just about all of the salvaged items used in our projects. Doors, windows, shutters, screens, drawers, and cabinets of all sizes and shapes are available. If you need two or more identical items—such as shutters for a folding screen—here's the place to buy them without a hassle. To get your project off the ground, see "Preparing a New Wood Surface" on page 12.

Tools and Supplies

$\mathscr{I}$n this section, we discuss the tools and supplies we used to prepare and paint the projects in this book. Our first list, "General Tools and Supplies," covers standard items and products. These are the items you will want to have on hand no matter which project you choose to work on. Our second list, "Technique-Specific Tools and Supplies," covers items that are particular to certain projects.

Painting Supplies

Tools for Prepping and Repairs

General Tools and Supplies

Shop for the items on this list at your local home improvement store, at craft and art supply stores, or by mail-order. (See "Suppliers" on page 94.) You may already have some of these items around the house.

AC's Acrylic Craft Paint Remover and a Scrubber Sponge: Use a scrubber sponge (a kitchen sponge with one rough surface) and AC's Acrylic Craft Paint Remover to remove dried acrylic paint from stencils.

Artist's Brushes and Stencil Brushes: Build your own assortment of flat, round, and angle-tipped artist's brushes, ¼" to 1½" wide. Artist's brushes yield smoother finishes than ordinary paintbrushes. They are perfect for applying sealers, primers, and base coats. They are also used in some decorative painting techniques. For stenciling, you'll need stencil brushes. For a closer look at brushes, see "Choosing Brushes" on page 8.

Artist's Water Basin: An artist's water basin serves as both a brush holder and a rinsing basin for your artist's brushes.

Basic Woodworking Tools: Keep a hammer, pliers, screwdrivers (straight-bladed and Phillips-head), an electric drill and drill bits, a palm-held sander, and a metal tape measure on hand. You'll use these tools for removing and reinstalling hardware, breaking out old glass, measuring, sanding, and other prepping and finishing tasks. Use the drill to make starter holes for your screws when you need to attach new hardware or join two pieces of wood together.

Brush Cleaner/Conditioner and Brush Scrubber: Use a brush cleaner/conditioner and a brush scrubber to clean stencil brushes. A cleaner/conditioner is better for your brushes than soap alone. There are many different types;

the product we use looks like a hard, white soap and comes in a tub. Brush scrubbers are small plastic implements with thin teeth that help remove paint from stencil-brush bristles.

Cotton Swabs: Cotton swabs work well for fixing small painting mistakes. While the paint is still wet, moisten a swab and use it to wipe away any unwanted paint. Swabs can also support wooden ball knobs as they are painted and allowed to dry. (See "Base Coat Painting" on page 12.)

Drop Cloth or Newspapers: Before you begin a project, protect your work surface with a drop cloth or several layers of newspaper.

Extender: Extender is a liquid that is mixed into acrylic paints to increase the amount of time that the paints stay wet. Adding a drop or two of extender to each color you use will make the paint more manageable and will help create smooth finished surfaces. Extender is especially useful when stenciling. It comes in 2-ounce bottles.

Goggles: Wear goggles whenever you are sanding to protect your eyes from fine dust particles. Also put on your goggles before breaking out old glass.

Heavy-Duty Canvas Gloves: Wear heavy-duty canvas gloves when breaking old glass or mirror out of salvaged doors and windows. The gloves can also offer general protection when you are hunting for pieces to salvage.

Painter's Tape and a Burnisher: Use removable painter's tape (also known as "blue tape") to affix stencils to your projects. When you peel off the tape, you won't remove your freshly painted finish along with it. You can also use blue tape to mask off areas that you don't want to paint. To form a tight seal, burnish or press down the edges of the tape with a burnishing tool or the edge of a credit card or any similar wooden or plastic item.

Paint Palette: You'll never dip a brush directly into a bottle of paint. Instead, you'll squeeze some paint onto a palette and load your brush

from it. A pad of disposable artist's palettes makes for easy cleanup. Just tear off the used sheet when you're though painting and throw it away. Disposable plastic plates make good substitutes.

Paint Pen: For signing your finished masterpiece, a fine-tipped black paint pen is convenient and easy to handle.

Paper Towels: Use folded paper towels to remove excess paint from stencil brushes and natural sea wool sponges (a special sponge used to create faux finishes). This paint removal process is known as "off-loading." You'll also need paper towels for your wet bags (see "Wet Bags" on page 8).

Primer: Before it is painted, a wood surface should be primed. We use water-based primer, since our acrylic paints are also water-based. A primer coat acts to seal the wood and helps subsequent layers of paint to bond properly.

Putty Knife: Use a putty knife to scrape out the old caulking around glass windowpanes.

Respirator or Dust Mask: Whenever you are sanding wood, wear a respirator or a dust mask and work in a well-ventilated area, preferably outdoors. Inhaling the airborne particles created during sanding isn't good for your health.

Sandpaper: Purchase sheets of sandpaper in several different grits, or degrees of roughness: coarse (40- to 60-grit), medium (80- to 120-grit), and fine (150- to 180-grit). You'll need the sandpaper to prepare wood surfaces before you paint them and to smooth dried coats of paint. Sandpaper is also used to create some worn effects.

Scissors: Keep a pair of scissors on hand for cutting cheesecloth or sandpaper. You can also use scissors to trim various project components, including lace, paper prints, and dried flower stems.

Tack Cloth and Lint-Free Rags: After sanding a wooden object for painting, use a tack cloth or lint-free rag to wipe away the residual dust. Rags also come in handy for wiping up spilled paint. Rags are used to create antique finishes, too.

Varnishes: Protect finished projects with several coats of water-based varnish. Use an interior varnish for projects that will remain indoors. Use an exterior varnish for projects that you plan to set outdoors.

Wet Bags: For stenciling, you'll need one brush for each paint color. As you switch back and forth between brushes, you'll place the paint-loaded brushes that you're not using into a "wet bag." A wet bag prevents the paint on a brush from drying out prematurely. To make a wet bag, moisten a paper towel, squeeze out the excess water, fold the towel, and place it in a plastic bag. Store your paint-dampened stencil brushes in the wet bag, with their bristles resting on the damp paper towel, until you are ready to use them again.

Wood Putty: Use wood putty to fill any cracks, dents, or holes in your project before painting it.

Wood Sealer: Apply a water-based wood sealer to unfinished wood to prevent warping. Give knots in the wood several coats of sealer. This will prevent pitch from bleeding out of the knots and through the primer and finishes you apply.

Technique-Specific Tools and Supplies

This list may seem long, but don't worry—you won't need everything all at once! Follow the "Tools and Supplies" list for your project to gather the items you will need. If you want to know more about a specific item, look it up alphabetically below. You can purchase these technique-specific tools and supplies at home improvement stores, at craft and art supply stores, and by mail-order. (See "Suppliers" on page 94.)

Acrylic Paints: Acrylic craft paints come in a wide range of colors, are easy to apply, and dry quickly. We used DecoArt Americana acrylics to paint the projects in this book. See the individual project instructions for the specific colors. You'll

Choosing Brushes

Your painting will always be at its best if you choose the right brush for the job. Here are descriptions of the artist's and stencil brushes we used to create the projects in this book.

Assorted Brushes

Flats: Flat brushes, with their squared-off bristles, are ideal for priming, base coat painting, blocking in large areas of color, and varnishing.

Angular Flats: The bristles of these flat brushes are trimmed at an angle. They're great for applying colors in tight corners and for painting one color next to another.

Long Liners: Also called script liner brushes, these brushes have very long bristles that come to a sharp point. These brushes are your best choice for painting fine lines.

Extreme Angulars: Use extreme angular brushes to cut in one color next to another color or to paint the spindles on posts and finials.

Stencil Brushes: You'll need several stencil brushes, ranging from ¼" to 1" in diameter. Unlike artist's brushes, stencil brushes are round and have stiff bristles and blunt ends.

need one 2-ounce bottle of each color, unless noted otherwise. Of course, you can also substitute your own color palette. Please note that acrylic colors dry darker than the color they are in their bottles. To test a color, brush some of the paint onto a piece of paper and let it dry.

Chalkboard Paint: Chalkboard paint can be applied with a brush or roller to many different surfaces. Once the surface is dry, the painted surface can be used just like a chalkboard.

Cheesecloth: Wadded-up pieces of cheesecloth are used to stipple glaze finishes.

Chip Brushes: Inexpensive chip brushes have rough stiff bristles that are wonderful for applying glazes.

Clear Silicone: Use clear silicone to set panes of glass and mirror into framed window and door openings.

Crackle Medium: This clear paint medium lets you create a crackled effect. (See "Crackle Finishes" on page 15.)

Decorative Accessories: This catchall category includes drawer pulls, doorknobs, faceplates, hooks, and wooden finials. Don't forget the hardware for attaching your pieces. For soft accents, you might use vintage lace, dried flowers, beaded fringe, art prints, and so forth. If you like a vintage look, shop for these items at antique shops and rummage sales.

E6000 Industrial Strength Adhesive: Use this adhesive to attach wooden ball knobs, bun feet, and other pieces that require extra-strong hold. Be sure to follow the manufacturer's instructions.

Embossing Tool: This crafter's tool is a stick with a small, hard metal ball at one or both ends. The balls are usually used to emboss paper, but you'll use them to create La De Da Dots. (See "La De Da Dots" on page 16.)

Etchall Etching Creme and Supplies: This glass-etching medium lets you create permanent frosted designs on a piece of glass. You'll also need Con-Tact paper, a pencil, and an X-Acto knife. Use spray glass cleaner and paper towels to clean the glass surface before you begin. A squeegee is handy for applying and removing the medium.

Eye Hooks, 18-Gauge Wire, and Picture Hangers: Use these picture-hanging accessories for wall-mounted projects.

Faux Glazing Medium: This pigment-free medium increases translucency. It can be used in several ways. For a sponged finish, mix the medium with acrylic paint. (See "Sponged Faux Finishes" on page 15.) For a negative glazed finish, mix it with gel stain. (See "Negative Glazed Finishes" on page 16.)

Foam Block or Egg Carton: A rigid foam block (used in floral crafts) or an egg carton becomes a handy drying stand for freshly painted wooden ball knobs. (See "Base Coat Painting" on page 12.)

Foam Core Board and Foam Core Knife: This lightweight, easy-to-cut board can be used as a mounting surface or a backing. The special foam core knife lets you make clean cuts, both straight and angled.

Friendly Woodworker or Advanced Woodworking Tools: A few of our projects require advanced woodworking skills and tools, such as jigsaws and table saws. You may wish to do what we did and recruit a local woodworker to tackle these parts of the project for you.

Gel Stain: Gel stains have several uses. You can brush them over painted and varnished surfaces and then wipe off the excess to give your project an antique, distressed appearance. Gel stain mixed with faux glazing medium lets you create a negative glazed finish. (See "Negative Glazed Finishes" on page 16.) Gel stains are available in 2-ounce bottles. We use DecoArt Americana Gel Stains.

Gesso: Gesso is a very thick, water-based artist's primer that yields an opaque, smooth finish. You'll need this special primer for the stenciling technique called "whiting out." (See the section on whiting out on page 14.)

Glass and Stained Glass: New glass insets can be cut to size to replace broken panes in recycled doors and windows. (See "Working with Glass" on page 11.)

Glass Brads: Glass brads are used with clear silicone to secure mirror panes in a window opening.

Glass Cutter and Lubricant: A glass cutter is used to cut glass, stained glass, and mirrors. You'll also need glass-cutting lubricant.

Gunther Glue: Use Gunther glue to affix glass mirror panels to a solid surface. There are several Gunther glue products from which to choose. We generally use Ultra/bond.

Hinges: Use hinges to join screen panels together and to attach windows to cabinets. There are many different hinge styles from which to choose. Don't forget screws.

Latex Gloves: We suggest that you wear latex gloves when working with glazes, as the process can be a little messy. Glove protection is a must when you work with etching cream.

Marking Pen: Use a permanent marking pen to mark cutting lines on glass.

Measuring Cup: You'll need a measuring cup to make some glaze and stain mixtures. Keep the cup with your painting supplies. Don't use it as a food container.

Metal Primer: Water-based metal primer is specially formulated for use on metal surfaces. It seals the surface and helps subsequent layers of paint to bond properly.

Mirror: Mirrored glass insets can be cut to size and glued directly over wood panels with Gunther glue. Choose a thin mirror—⅛" to ¼" thick—so that the project does not become too heavy.

Mixing Tubs and Stirring Sticks: Mixing tubs and a few stirring sticks will come in handy when you need to mix glazes.

Natural Sea Wool Sponges: Use natural sea wool sponges to apply sponged faux finishes. Available at craft stores, these sponges are much softer than ordinary sea sponges. (See "Sponged Faux Finishes" on page 15.)

Packaging Tape: Use wide plastic packaging tape to secure lightweight foam core board to the backs of some projects for a more finished look.

Pine Lumber and ¼" Plywood: Some projects require additional wood components. We used pine lumber and ¼"-thick plywood to make the pieces we needed.

Spray Adhesive: Use a low-tack spray adhesive for lightweight applications such as mounting prints and attaching lace. Use industrial spray adhesive to mount boards to screens.

Staple Gun and Staples: Use a staple gun to secure lightweight pieces, such as foam core board or a beaded fringe, to a wood backing.

Stencils: A stencil is a plastic sheet with a cutout design. Paint is applied with a stencil brush through the cutout openings to add the design to your project. Stencils can be used again and again, giving you a lot of design impact with little effort. We used precut Stencilled Garden stencils to create the projects in this book. The project instructions list the specific designs and paint colors we used. Of course, you can substitute other designs. Stencils are usually packaged in plastic bags. Save the bags; they'll help protect your stencils when you're not using them.

Straight Edge Metal Ruler: Use a rigid metal ruler when marking cutting lines on glass and as a guide for your glass cutter.

Watercolor Pencils: Watercolor pencils are ideal for marking off areas to be painted, such as the corners on the "Reflection in Time" project on page 33. Light-colored pencil lines won't show through the paint that covers them.

White Tapered Candle: A candle will help you create the look of multiple layers of peeled paint. Rub candle wax onto a dry, painted surface. Paint over it with the new color and let dry. The wax makes it easier to sand off the top coat to reveal the color underneath.

Wooden Ball Knobs: A wooden ball knob is a round ball with one flat surface and a hole running partway into its center. We attach these knobs to many projects to serve as bun feet. They come in several different sizes. On larger projects, we use actual bun feet.

Wood Screws: Use wood screws to join two pieces of wood together. Wood screws come in different lengths and diameters; you'll need to choose a size that's appropriate for your project. It helps to drill a small starter hole in the wood beforehand. Glue is sometimes used in addition to screws for an extra-strong join.

Surface Preparation

The first step in any painted wood project is to prepare the surface. If you skip this step, the paint you apply won't adhere properly. Before you begin, protect your floor or work surface with a drop cloth or several layers of newspaper. Gather all your tools and supplies. If your project has existing hardware, remove it and set it aside to be reinstalled (or replaced) later.

Preparing an Old Wood Surface

Sanding will usually remove loose particles and old paint finishes. At the same time, you'll be roughening the surface of any finish that remains. This roughening up is important because new paint won't adhere to a slick finished surface. Start with coarse-grit sandpaper and change to finer grits as the work progresses. Remember to wear goggles and a respirator or dust mask. If possible, do your sanding outdoors; you'll be generating a lot of dust. When you're finished sanding, wipe the project with a tack cloth or rags to remove any dust.

If your project is rickety, repair it or have a woodworker repair it for you. Use wood putty to fill any holes, pits, or dents that are marring the project surface. Allow the putty to dry. Then sand the area lightly to make the putty filler level with the surrounding wood.

The next step is priming. Use a water-based primer and a flat artist's brush, 1" to 1½" wide. Apply several thin, even coats of primer to the entire project. For the smoothest possible finish, keep your brush strokes going in the same direction. After the primer has dried, sand it lightly with your finest-grit sandpaper. Then wipe the project with a tack cloth or rags.

Working with Glass

To remove old or broken glass: Be sure to wear goggles and heavy-duty canvas work gloves. Rest the wooden item across the rim of a large garbage container and break out the glass with a hammer. Use a putty knife to loosen stubborn shards and scrape out the old dried-up caulk. Use pliers to pull out old glass brads.

To cut replacement glass: Lay the glass or mirror on a flat work surface. (Make sure the work surface is absolutely flat, to avoid stressing the glass.) Use a permanent marking pen and a ruler to mark the size and shape of the replacement pane on the surface of the glass. Align a metal ruler on a marked line. Beginning ⅛" from the edge of the glass, draw the glass cutter along the ruler with even pressure to score a line. It helps to work from a standing position and to lubricate the glass cutter with glass-cutting lubricant before you begin. Score the glass only once; never go over the same score line twice.

To complete the cut, position the glass so that the score line falls slightly beyond and parallel to the edge of the work surface. Hold down the glass firmly with one hand. Use the other hand (wear a work glove) to snap off the overhanging section with a sharp downward movement. Cutting glass isn't difficult, but you may prefer to have a glazier cut your pieces for you.

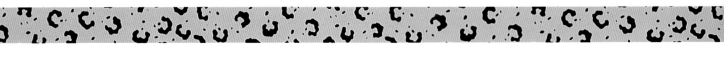

Preparing a New Wood Surface

You won't always be working with old wood. Some of our projects use a combination of old and new wood. Other times, you may find it more convenient to start with a new wood piece instead of recycling something old.

Sand the surface with a medium- to fine-grit sandpaper. Work outdoors if you can, and remember to wear goggles and a dust mask or respirator. Always sand in the same direction as the wood grain. If the wood surface has dents, cracks, or nail holes in it, fill them with wood putty, allow the putty to dry, and sand it smooth. Sand lightly to remove traces of glue and to smooth all surfaces. Wipe the project with a tack cloth or rags when you're finished.

The next step is sealing. Use a wood sealer and a flat artist's brush, ¼" to 1½" wide. Apply a thin, even coat of wood sealer to the entire project surface. The sealer will help prevent warping and will also raise the wood grain slightly. Allow the sealer to dry. Then sand the surface very lightly with fine-grit sandpaper and wipe with a tack cloth or rag. If the wood has any knots, seal them several times more, drying and sanding lightly between applications. Sealing the knots will prevent pitch from bleeding out of knots and through the painted finishes.

After sealing comes priming. Follow the same priming method as for an old wood surface.

Base Coat Painting

Once you have finished the surface preparation part of your project, you'll be ready to paint the base coat. The base coat is the foundation for all the decorative painting that follows.

Read the individual project instructions to learn which base coat colors to use and where to apply them. Pour a little paint onto a paint palette and work some of it into a flat artist's brush. Brush the paint onto the project, always stroking in the same direction. If the paint won't brush on smoothly, dip your brush into water to moisten it slightly before working the paint into the bristles.

Allow the first coat of paint to dry. If the painted surface is rough, use your finest-grit sandpaper to smooth it, but be careful to remove as little paint as possible. Wipe away any dust. Apply as many coats as necessary to achieve smooth, opaque coverage, letting each coat dry before applying the next.

To paint a wooden ball knob, insert one end of a cotton swab into the hole in the knob. Use the other end of the swab as a handle so that you don't have to touch the knob itself. When you finish sealing, priming, or painting the knob, insert the free end of the swab into a foam block to hold the knob upright for drying. You can also prop the knob and its swab handle in an empty egg carton for drying.

Decorative Painting

This section provides complete instructions for all the decorative painting techniques featured in this book. You don't need to memorize every detail; just turn back to these pages whenever you need to refresh your memory.

Stenciling

Stenciling is a remarkably easy painting technique. If you've never tried it before, we suggest practicing on paper before tackling your project. We used stencils from the Stencilled Garden for all of our projects. These stencils are available at specialty stencil shops and from mail-order suppliers. (See "Suppliers" on page 94.) Each set of project instructions specifies which stencils to use and which colors to apply. Of course, you may also choose to substitute other stencil designs.

To position the stencil: Place your stencil on the project, locating the open design cutouts over the areas where you'd like the painted designs to appear. Then tape the outer edges of the stencil to the project with removable painter's tape.

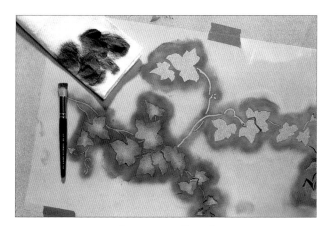

Positioning a Stencil

To load and off-load the stencil brush: Pour a little paint onto a paint palette, add one or two drops of extender, and use the handle end of your stencil brush to mix in the extender well. Next, holding your stencil brush straight up, pick up a small amount of paint with the tips of the brush bristles. Work the paint into the bristles by swirling them in a circular motion on a clean section of the palette.

One trick to successful stenciling is to have only the tiniest amount of paint on your brush bristles. To off-load the excess paint, hold the brush upright and with a firm, circular motion, rub the bristles on a folded paper towel. Then, on a clean portion of the paper towel, wipe the brush in an X motion to remove excess paint from the outer bristles.

Working Paint into the Brush

Off-loading the Brush

To apply the paint: For a smooth stenciled surface, "swirling" is the best technique to use. Hold your stencil brush perpendicular to the project surface and apply the paint by moving the bristle ends in tiny circles. To add texture and depth to a stenciled design, hold the brush perpendicular to the surface, but instead of swirling the bristles, dab the brush straight up and down—a process known as "stippling."

To create shading in the stenciled design, first create sharp, crisp edges by applying paint lightly all the way around the outer edges of the design opening. As you work, blend paint into the interior design area but apply less pressure to the brush. You will notice that the color appears lighter toward the center of the design. By varying the pressure you apply to your brush and the amount of time you spend stenciling a given area, you can achieve a wide range of values with a single color.

Shading with One Color

For added contrast within a design, use more than one color. Let each color dry before applying the next, use a different brush for each color, and leave the stencil in place until you've applied all the colors. (Once you've removed a stencil, it's very difficult to realign it in exactly the same position.)

Shading with Two Colors

To white out: Before you can stencil on top of a black or dark-colored base coat, you must "white out" the areas where your stenciled designs will appear. Otherwise, your stenciled colors won't show up properly.

Whiting out is easy. Position your stencil on the project and secure it with tape. Use a stencil brush to apply gesso through the stencil. Leave the stencil in place and allow the gesso to dry. (If you like, you may use a blow dryer to speed up the drying process.) Then stencil your design with the desired colors directly over the white gesso coat. Never remove the stencil until you've applied all the colors.

To clean and store stencils and brushes: After using a stencil, you should clean it as soon as possible. The longer the paint remains on the stencil, the harder it will be to remove. Unfortunately, cleaning is hard on stencils and makes it all too easy to damage them.

One way to clean stencils is with AC's Acrylic Craft Paint Remover. (See "Suppliers" on page 94.) Place the stencil in a sink. Pour the cleaner over it, let everything sit for a minute or two, and then gently scrub off the paint with a scrubber sponge. Rinse the stencil with hot water to remove the cleaner.

Another way to clean stencils is to use hot water and a bit of elbow grease. Place the stencil under hot running water and rub it gently with the rough side of a scrubber sponge.

To dry a stencil after cleaning it, place it on a towel and either let it air-dry or pat it dry with a paper towel. Store clean stencils in their original plastic bags, stacking the bags so that the stencils lie flat.

To clean a stencil brush, first moisten the bristles under running water and scrub them over the surface of a brush scrubber. Next, apply brush cleaner/conditioner to the bristles and scrub them over the brush scrubber again. Rinse the bristles under water. Repeat until the suds are clear and colorless. Squeeze out the excess water and place the brush on its side to dry.

Sponged Faux Finishes

A sponged faux finish is made by applying a glaze mixture on top of a base coat. The texture of the sponge creates variegated textures and colors. To make the glaze mixture, put paint and faux glazing medium in a mixing tub in the proportions recommended in the project instructions. Use a stirring stick to mix them together.

Put on latex gloves. Dampen a natural sea wool sponge with water, wring out the excess moisture, and dip the sponge into the glaze mixture. To work the mixture into the sponge, rub the sponge in a circular motion on a clean section of your palette. Then dab the sponge onto a paper towel to off-load some of the glaze mixture.

Using a light touch and a stippling motion, sponge the mixture onto the project. Allow the sponged glaze to dry thoroughly. If you like, you may apply more coats, letting each one dry before applying the next.

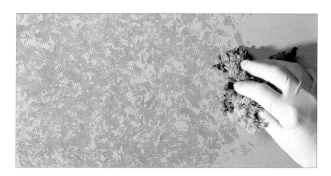

Sponged Faux Finish

Antique Finishes

For a distressed or antique look, apply one coat of varnish over an opaque base coat. Allow the varnish to dry. Then use an artist's brush to apply gel stain over the varnish, brushing in the same direction as you did when applying the base coat. While the gel is still wet, use a rag to wipe it. As you wipe, the base coat colors will darken and acquire an aged look. For a darker finish, apply several coats of the gel stain, wiping and drying after each application.

Antique Finishes

Crackle Finishes

In a crackle finish, the top coat of paint develops cracks that allow the base coat color to show through. Begin by applying the base coat to the project, as many layers as are needed for smooth, even coverage. When the final base coat is dry, pour some crackle medium onto your palette. Using an artist's brush, spread the medium over the area that you want to crackle. For small cracks and a subtle finished texture, apply a thin coat. For larger, more pronounced cracks, apply a thicker coat. Allow the medium to dry.

Next, brush the top coat paint color over the crackle medium. The stronger the contrast between the top coat and base coat colors, the more noticeable the finished crackle effect will be. Use long, even strokes, running your brush in one direction only, one stroke next to another. Once you have painted an area, do not return to it. The top coat will begin reacting with the crackle medium almost immediately, and if you disturb it, your brush will lift the top coat right off the project surface. There's no way to repair a botched crackle job, short of sanding your project and starting over.

Crackle Finishes

Negative Glazed Finishes

Negative glazed finishes are applied over a base coat and/or stenciled artwork to create aged textures and colors. Start by pouring some faux glazing medium into a mixing tub. Add a little gel stain and mix with a stirring stick. Continue adding and mixing in gel stain, a little at a time, until you achieve the desired color.

Use a chip brush to apply the glaze mixture to your project. To create the textured finish and also to eliminate brush strokes, wad up a piece of cheesecloth and "pounce" it over the glazed surface, removing the excess glaze. Allow the finish to dry thoroughly. If the first application of glaze isn't dark enough to suit you, or if you want to "age" the edges or corners by making them a little darker, apply more glaze coats where desired. Be sure to let each coat dry before applying the next one.

Negative Glazed Finish

La De Da Dots

La De Da Dots are simply raised dots of paint. To create them, pour a little paint onto your palette. Touch one end of an embossing tool to the paint, picking up a small amount. Then touch the tool to the project to transfer the color. Repeat this process for every dot. The secret to dots that are raised and not flat is to pick up fresh paint from the palette for each dot that you make.

Beginners sometimes have trouble spacing their dots, bunching them up in some areas and scattering them too far apart in others. To prevent these problems, visualize a small triangle in the middle of the area to be painted. Apply a dot to each point of the imaginary triangle. Now visualize another triangle linked to the first one. Apply dots to the points of this second triangle. If you continue applying dots in this way, you'll find that they are evenly spaced.

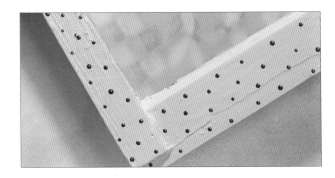

La De Da Dots. Visualize triangles to space the dots.

Final Touches

We've always found it fun to personalize our projects by signing and dating each piece. When Judy's grandson, Josh, was very young and tried to say Judy, out would come "Jubee" instead. Judy began signing her work this way, writing the letters "Ju" followed by a stenciled bee. Jennifer signs each piece differently. If the piece is a gift to someone, she often writes a personal message for the recipient. At other times, she just signs "J. Ferguson" and the date.

The very last step is applying varnish. Varnish is very important. The coats of varnish protect all the work you've put into your masterpiece. Use a water-based varnish and apply it with a flat artist's brush. To avoid drips and runs, apply at least three thin, even coats. Allow each coat to dry thoroughly before applying the next.

Sign and date each project.

Projects

Doors

We could write an entire book simply on the art of recycling doors—there are so many different styles to work with. We look for doors with lots of raised panels and moldings to add architectural interest and character to our projects. We've turned old doors into dressing screens, room dividers, hall trees, and coffee tables. Of course, if you find a really great-looking door, you could simply hang it on the wall as art.

Windows

Windows come in all shapes and sizes. When they are no longer needed to frame outdoor views, windows can focus attention on other objects. An old window, revived and painted, is a clever way to showcase your treasures, frame a favorite print, or protect delicate dried flowers. The glass panes can be replaced with stained glass or mirrors to create unique and useful home accents. Hang one of your transformed windows on a blank wall, the same way you would a picture, for an instant architectural remodeling.

Mesh Screens

Mesh screens remind us of warm, drowsy days in late summer when bees are buzzing and the garden is bursting with produce. We find that old screened doors and windows are easy to come by. Since they don't have any heavy glass, they're also easy to tote away. In addition to painting the wooden frame, we put designs on the mesh itself. Both sides of the mesh can be painted, and the designs don't have to match. Try that idea on your own swinging screen door for a different look coming and going.

Cabinet Doors

A cupboard door is like a blank canvas. There are so many different sizes and shapes, we'll never run out of ideas. With a little chalkboard paint and some cork, we came up with a practical kitchen message center. Doors with raised panel edges, we discovered, make handsome mirror frames. Small, paneled doors seem to call out to us, asking us to transform them into something special. We know what one of these little doors wants to be almost as soon as we see it. Spend a little time with a cupboard door, let it speak to you, and see what you come up with.

Shutters

Originally used to close up a house before a storm, the window and door shutters on today's homes are more decorative than functional. It really doesn't matter to us where our shutters "lived" in the past. We just love all those slats for holding mail and notes, artwork and postcards, even CDs. Hinge a few shutters together for an instant privacy screen, a room divider, or a converstion piece for a neglected corner.

Drawers

A trashformation drawer is so much more than a drawer. Shallow drawers make useful serving trays—just add handles at each end. We've turned drawers upside down and even pulled off a front panel or two to use separately. Our idea to join two drawers for an under-the bed storage container made a big hit with our fimilies. All you need is a creative eye and a little imagination.

Check Your Coat and Hat
Hall Tree

Judy's primary business is salvaging usable parts of older homes. The wonderful old door and tool storage box for this project came from a home that was scheduled to be torn down. Before the demolition began, the owners gave Judy the opportunity to "strip" the house for pieces she could use.

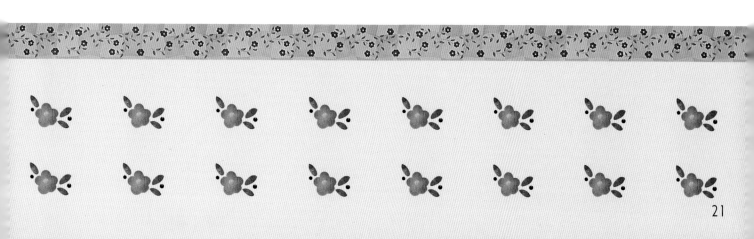

Tools and Supplies

1 paneled wooden door (ours is 30" x 79")

1 storage box (ours is 19" x 30" x 20¼")

4 wooden bun feet, 3" in diameter

Mirror

1 decorative doorknob and faceplate

3 decorative hooks

Wood screws

Acrylic paints (see below)

Gesso

Gunther glue (we generally use Ultra/bond)

E6000 adhesive

Embossing tool

Glass cutter

Marking pen

Stencils (see below)

Straight edge metal ruler

General tools and supplies

Paints and Stencils

We used the following paints and stencils to create our "Check Your Coat and Hat" project. (See "Suppliers" on page 94.) For a different look, substitute the colors or stencils of your choice.

DecoArt Americana acrylic paints:

Limeade (DA206)—10 bottles

Black (DA67)

Jade Green (DA57)

Plum (DA175)

Black Plum (DA172)

Hauser Light Green (DA131)

Plantation Pine (DA113)

French Vanilla (DA184)

Moon Yellow (DA07)

Marigold (DA194)

Easy Blend Charcoal Grey (DEB28)

Stencilled Garden stencils:

Gingham (TSG112L)

Folk Art Posies (TSG241)

Hungry Kitty (TSG720)

Wild Posies (TSG190)

Whimsey Critters (TSG828)

Cherry Pickin (TSG709L)

Buzzy Bee Border (TSG825)

Stencil Color Guide

Gingham: Jade Green

Folk Art Posies: Plum, Black Plum, Hauser Light Green, Plantation Pine

Hungry Kitty: French Vanilla, Moon Yellow

Flowers from Wild Posies: Plum, Black Plum, Hauser Light Green, Plantation Pine

Whimsey Critters: Limeade, Jade Green, Plum

Cherry Pickin: Plum, Black Plum, Hauser Light Green, Plantation Pine

Bees from Buzzy Bee Border: Moon Yellow, Marigold, Black, Easy Blend Charcoal Grey

Instructions

1 Prepare the wooden door and storage box for painting. (See "Surface Preparation" on page 11.)

2 Paint the door, the body of the storage box, and the bun feet with Limeade. Paint the storage box trim and lid with Black. For the correct color placement, refer to the project photo. Apply as many base coats of paint as necessary to achieve smooth, opaque coverage. (See "Base Coat Painting" on page 12.)

3 Use gesso to white out areas on the Black base coat that you plan to stencil. (See the section on whiting out on page 14.)

④ Stencil the designs, referring to the project photo on page 20 and the "Stencil Color Guide" on page 22 for placement. (For general instructions, see "Stenciling" on page 13.) To make the kitty "overlap" the bowl on the storage lid, stencil the kitty first. After the paint has dried, shield the stenciled kitty by placing the solid, positive (or "fallout") portion of the stencil over it. Position the bowl stencil on top and stencil the bowl. Using the fallout will make the bowl appear to be under the kitty. To create shadows around the kitty, leave the fallout in place and stencil around the edge with Easy Blend Charcoal Grey.

⑤ Use an embossing tool to apply French Vanilla La De Da Dots to the flower centers, Plum La De Da Dots around the rim of the bowl, and Black La De Da Dots randomly around the flowers on the kitty and to the bun feet. (See "La De Da Dots" on page 16.)

⑥ Sign your project. Allow the paint to dry for several days. Apply at least three coats of varnish to protect your work. (See "Final Touches" on page 17.)

⑦ Test-fit the door against the storage box. Measure the door panels that will be visible after the wood pieces are joined. Use a marking pen, metal ruler, and glass cutter to cut one piece of mirror for each panel. (See "Working with Glass" on page 11.) Install the mirror panels with Gunther glue, following the manufacturer's instructions and recommended cure time.

⑧ Join the storage box and door using wood screws. Use E6000 adhesive to attach the knob "feet" to the bottom of the storage box.

⑨ Install the doorknob, faceplate, and decorative hooks on the door as shown.

CHECK YOUR COAT AND HAT HALL TREE

Here's Looking at You
Dressing Screen

One day, on one of our junking trips in our own backyard, we met a gentleman who dealt in salvaged home items. He became a wonderful resource, as he had many opportunities to get salvaged pieces that we could use. He acquired these doors—one of several sets that we've received through him over the years—from an old farmhouse.

Tools and Supplies

3 wooden doors (ours are 80" tall and 21½" to 24" wide)

Mirror

4 door hinges

1 doorknob and faceplate

Acrylic paints (see below)

Faux glazing medium

Gel Stain (DS29)

Gunther glue (we generally use Ultra/bond)

Cheesecloth

3" chip brush

Glass cutter

Marking pen

Measuring cup

2-quart mixing tub and stirring stick

Natural sea wool sponge

Stencils (see below)

Straight-edge metal ruler

General tools and supplies

Paints and Stencils

We used the following paints and stencils to create our "Here's Looking at You" project. (See "Suppliers" on page 94.) For a different look, substitute the colors or stencils of your choice.

DecoArt Americana acrylic paints:

Buttermilk (DA03)—12 bottles

Khaki Tan (DA173)

French Mocha (DA188)

Antique White (DA58)

Antique Maroon (DA160)

Stencilled Garden stencils:

Villa Francisco Scroll (TSG527)

Décor Border (TSG532S)

Stencil Color Guide

Villa Francisco Scroll: Antique White, Khaki Tan, French Mocha, Antique Maroon

Décor Border: Antique White, Antique Maroon

Instructions

1. Prepare the wooden doors for painting. (See "Surface Preparation" on page 11.)

2. Paint each door with Buttermilk. Apply as many base coats of paint as necessary to achieve smooth, opaque coverage. (See "Base Coat Painting" on page 12.)

3. On your palette, mix one part Khaki Tan paint and one part faux glazing medium. Moisten a natural sea wool sponge with the glazing mixture. Lightly sponge the entire surface of each door, reloading the sponge as needed. (See "Sponged Faux Finishes" on page 15.) Let dry completely.

4. Stencil the designs, referring to the project photo on page 24 and the "Stencil Color Guide" above for placement. (For detailed stenciling instructions, see "Stenciling" on page 13.) Let dry.

5. Use medium-grit sandpaper to lightly sand all the stenciled areas and create an aged appearance.

6. Apply one coat of varnish to the doors. Allow the varnish to dry completely.

7. Use a measuring cup to measure and pour 16 ounces of faux glazing medium into the mixing tub. Add 4 ounces of gel stain and mix with a stirring stick. Use a chip brush to spread the glaze mixture over a small section of the varnished area. Dab a wadded-up piece

of cheesecloth over the surface to eliminate the brush strokes and stipple the glaze. Continue applying and stippling the glaze, small areas at a time, over the entire varnished surface. (See "Negative Glazed Finishes" on page 16.) Allow the glaze to dry completely.

8 Sign your project. Allow the paint to dry for several days. Apply at least three coats of varnish to protect your work. (See "Final Touches" on page 17.)

9 Measure the inside panel of each door. Use a marking pen, metal ruler, and glass cutter to cut one piece of mirror for each panel. (See "Working with Glass" on page 11.) Install the mirror panels using Gunther glue, following the manufacturer's instructions and recommended cure time.

10 Install the doorknob and faceplate on one door. Hinge the doors together.

Botanical Garden
Room Divider

When you are in the salvage business, it's worth your while to stay in contact with local real estate brokers, developers, and demolition companies. One particular relationship we cultivated led to the rescue of these doors. Not only did Judy have the opportunity to salvage all possible items from this home, but she also had the joy of working side by side with her grandson Josh, who helped strip the house and give these items a second chance.

Tools and Supplies

1 French door (ours is 24" x 78")

1 three-paneled door (ours is 24" x 79½")

1 decorative doorknob and faceplate

2 hinges

Stained glass

Glass brads

Acrylic paints (see below)

Gel Stain (DS30)

Clear silicone

Glass cutter

Marking pen

Stencils (see below)

Straight edge metal ruler

General tools and supplies

Paints and Stencils

We used the following paints and stencils to create our "Botanical Garden" project. (See "Suppliers" on page 94.) For a different look, substitute the colors or stencils of your choice.

DecoArt Americana acrylic paints:

Light Buttermilk (DA164)—10 bottles

White (DA01)

French Mauve (DA186)

Raspberry (DA28)

Jade Green (DA57)

Plantation Pine (DA113)

Santa Red (DA170)

Napa Red (DA165)

Dusty Rose (DA25)

Gooseberry Pink (DA27)

Marigold (DA194)

Country Blue (DA41)

Prussian Blue (DA138)

Hauser Light Green (DA131)

Asphaltum (DA180)

Stencilled Garden stencils:

Calla Lilly (TSG609)

Morgan's Tulips (TSG611)

Peggy's Poppies (TSG610)

Blueberry Vine (TSG406)

Stencil Color Guide

Calla Lily: White, French Mauve, Raspberry, Jade Green, Plantation Pine

Morgan's Tulips: Santa Red, Napa Red, Jade Green, Plantation Pine

Peggy's Poppies: Dusty Rose, Gooseberry Pink, Jade Green, Plantation Pine, Marigold

Blueberry Vine: Country Blue, Prussian Blue, Hauser Light Green, Plantation Pine, Asphaltum, White

Instructions

1 Use a hammer to break out the existing glass, if any, from your French door. (See "Working with Glass" on page 11.)

2 Prepare both wooden doors for painting. (See "Surface Preparation" on page 11.)

3 Paint both doors with Light Buttermilk. Apply as many base coats of paint as necessary to achieve smooth, opaque coverage. (See "Base Coat Painting" on page 12.)

4 Stencil the designs, referring to the project photo on page 28 and the "Stencil Color Guide" above for placement. (For detailed instructions, see "Stenciling" on page 13.)

5 Apply one coat of varnish to the doors. Allow the varnish to dry completely.

6. Use a 1" artist's brush to apply gel stain over the varnished doors. It will appear streaky and uneven. Wipe the wet gel stain with a rag to create an antique look. (See "Antique Finishes" on page 15.) Allow the surface to dry completely.

7. Sign your project. Allow the paint to dry for several days. Apply at least three coats of varnish to protect your work. (See "Final Touches" on page 17.)

8. Measure the openings on the French door. Use a marking pen, metal ruler, and glass cutter to cut a piece of stained glass for each opening. (See "Working with Glass" on page 11.) Place the stained glass panes in the openings. Apply silicone along the outside edge of each pane. Allow to dry, following the manufacturer's recommendations.

9. Install the doorknob and faceplate on the paneled door. Clean and reattach any other decorative hardware. Hinge the doors together.

Reflection in Time
Wall Mirror

This window was one of many that came from a church that was replacing its old windows. Judy was approached by the contractor and ended up buying all the windows from the church.

Tools and Supplies

Six-paned window frame (ours is 24" x 30")

Mirror

3 decorative drawer knobs

2 eye hooks

18-gauge wire

Glass brads

Acrylic paints (see below)

Clear silicone

Embossing tool

Glass cutter

Marking pen

Stencils (see below)

Straight edge metal ruler

Watercolor pencil (optional)

General tools and supplies

Paints and Stencils

We used the following paints and stencils to create our "Reflection in Time" project. (See "Suppliers" on page 94.) For a different look, substitute the colors or stencils of your choice.

DecoArt Americana acrylic paints:

Light Buttermilk (DA164)—2 bottles

Soft Lilac (DA237)

Santa Red (DA170)

Hauser Light Green (DA131)

Napa Red (DA165)

Plantation Pine (DA113)

Black (DA67)

Country Blue (DA41)

Stencilled Garden stencils:

Cherries Jubillee (TSG184)

Girly's Flowers (TSG175)

Squiggles & Dots (TSG178)

Little Checks (TSG707)

Checkerboards (TSG706)

Stencil Color Guide

Cherries Jubillee: Santa Red, Napa Red, Hauser Light Green, Plantation Pine

Girly's Flowers: Soft Lilac, Country Blue, Hauser Light Green, Plantation Pine

Squiggles & Dots: Black

Little Checks: Black

Checkerboards: Black

Instructions

1. Use a hammer to break out the existing glass, if any, from your window frame. (See "Working with Glass" on page 11.)

2. Prepare the window frame for painting. (See "Surface Preparation" on page 11).

3. Paint the window frame with Light Buttermilk. Apply as many base coats of paint as necessary to achieve smooth, opaque coverage. (See "Base Coat Painting" on page 12).

4. Use painter's tape to diagonally mask off the two top corners of the window frame. Also mask off the front bottom section. Burnish the tape edges. Paint the masked-off sections with Soft Lilac as shown in the project photo. Allow the paint to dry before removing the tape. (These sections can also be painted freehand, if you prefer. Use a watercolor pencil to draw faint guidelines.)

5 Stencil the designs, referring to the project photo on page 32 and the "Stencil Color Guide" on page 34 for placement. (For general instructions, see "Stenciling" on page 13.)

6 Use an embossing tool to apply Black La De Da Dots randomly around the flowers and cherries and along the diagonal lines at the two corners. Apply Santa Red La De Da Dots randomly around the squiggles on the corners. (See "La De Da Dots" on page 16.)

7 Sign your project. Allow the paint to dry for several days. Apply at least three coats of varnish to protect your work. (See "Final Touches" on page 17.)

8 Measure each of the six window openings. Use a marking pen, metal ruler, and glass cutter to cut one piece of mirror for each opening. (See "Working with Glass" on page 11.) Install the mirror panels into the openings and secure with glass brads. Apply silicone along the outside edge of each pane. Allow to dry, following the manufacturer's recommendations.

9 Attach the eye hooks and 18-gauge wire to back of window frame for hanging. Attach decorative knobs to the front bottom section.

Stained Glass Print
Wall Art

We cherish our trips together. One of our favorite trips is to our publisher. Our road trip north to Washington State offers many opportunities for junking, and we have our favorite stopping places along the way. We found this particular window at a salvage yard in Oregon.

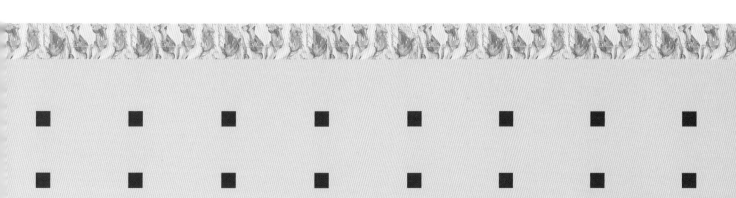

Tools and Supplies

1 three-pane window frame (ours is 29" x 48")

Art print

Stained glass

Clear glass

Glass brads

2 eye hooks

18-gauge wire

Foam core board

Packaging tape

Acrylic paints (see below)

Gel Stain (DS30)

Clear silicone

Spray adhesive

Foam core knife

Glass cutter

Marking pen

Straight edge metal ruler

White tapered candle

General tools and supplies

Paints

We used the following paints to create our "Stained Glass Print" project. (See "Suppliers" on page 94.) For a different look, substitute the colors of your choice.

DecoArt Americana acrylic paints:

   Terra Cotta (DA62)

   Sand (DA04)

Instructions

1. Use a hammer to break out the existing glass, if any, from your window frame. (See "Working with Glass" on page 11.)

2. Prepare the window frame for painting. (See "Surface Preparation" on page 11.)

3. Paint the window frame with Terra Cotta. Apply as many base coats of paint as necessary to achieve smooth, opaque coverage. (See "Base Coat Painting" on page 12).

4. To create a peeled paint look, rub the candle over the window frame in selected areas, pressing hard so that the wax transfers and sticks. Apply two coats of Sand paint to the entire piece, allowing each coat to dry completely. To expose the Terra Cotta base coat, sand the surface with 100-grit sandpaper, concentrating in the areas where you applied the wax. Expose as much of the base coat color as you desire. The wax prevents the top coats of paint from sticking and makes the sanding process easier. Use a tack cloth to wipe off the dust.

5. Use a flat artist's brush to apply one coat of varnish to the piece. Allow the varnish to dry completely. Use a 1" artist's brush to apply gel stain to the varnished window frame. The stain will appear streaky and uneven. Wipe the wet gel stain with a rag to create an antique look. (See "Antique Finishes" on page 15.) Allow the surface to dry completely.

6. Sign your project. Allow the paint to dry for several days. Apply at least three coats of varnish to protect your work. (See "Final Touches" on page 17.)

7. Measure the openings in your window frame. Use a marking pen, metal ruler, and glass cutter to cut a clear glass panel for the "print" opening and stained glass panels for the other openings. (See "Working with Glass" on page 11.) Install the glass into the openings. Apply silicone along the outside edge of each pane. Allow to dry, following the manufacturer's recommendations.

8. Use a foam core knife to cut a piece of foam core board to fit behind the clear glass pane. Trim the art print to the same size. Use spray adhesive to mount the print on the foam core board. Position the print behind the glass and seal the back outside edge with packaging tape.

9. Attach eye hooks and 18-gauge wire to the back for hanging.

A Window of Treasures
Shadow Box

We believe everything in life deserves a second chance—even unwanted windows. A local window company has started giving us the old windows it replaces. We do our part by recycling them into wonderful works of art. Our landfills are full enough already. This salvaged beauty has a new life as a shadow box.

Tools and Supplies

1 window frame (ours is 28" x 29")

1 decorative doorknob and faceplate

Glass

Pine lumber (see step 2)

¼" plywood

Piano hinge

2 heavy-duty picture hangers

Acrylic paints (see below)

Clear silicone

Friendly woodworker or advanced woodworking tools

Embossing tool

Glass cutter

Marking pen

Stencils (see below)

Straight edge metal ruler

General tools and supplies

Paints and Stencils

We used the following paints and stencils to create our "Window of Treasures" project. (See "Suppliers" on page 94.) For a different look, substitute the colors or stencils of your choice.

DecoArt Americana acrylic paints:

Soft Sage (DA207)—3 bottles

Wisteria (DA211)

Violet Haze (DA197)

Hauser Light Green (DA131)

Plantation Pine (DA113)

Moon Yellow (DA07)

Marigold (DA194)

Black (DA67)

Easy Blend Charcoal Grey (DEB28)

Stencilled Garden stencils:

Diamond Vines (TSG242L)

Wild Posies (TSG190)

Honey Pot (TSG193)

Stencil Color Guide

Diamond Vines: Hauser Light Green, Plantation Pine, Wisteria, Violet Haze

Flowers from Wild Posies: Hauser Light Green, Plantation Pine, Wisteria, Violet Haze

Bees from Honey Pot: Moon Yellow, Marigold, Black, Easy Blend Charcoal Grey

Instructions

1 Use a hammer to break out the existing glass, if any, from your window frame. (See "Working with Glass" on page 11.)

2 Measure the window frame to determine the width and length of the cabinet. The depth of the cabinet is 5½". You will need 2 sides, 1 top, 1 bottom, and 1 shelf, all cut from pine; and 1 back, cut from ¼" plywood. Calculate the dimensions for each piece, factoring in the wood thickness, and cut the pieces with a table saw. Assemble the cabinet as shown; leave out the shelf, which will be installed after stenciling. (We recruited a woodworker to build the cabinet for us.)

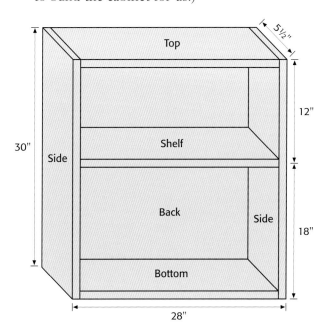

Cabinet Assembly

③ Prepare the window frame, cabinet, and shelf for painting. (See "Surface Preparation" on page 11.)

④ Paint all the wood pieces with Soft Sage. Apply as many base coats of paint as necessary to achieve smooth, opaque coverage. (See "Base Coat Painting" on page 12).

⑤ Stencil the designs, referring to the project photo on page 40 and the "Stencil Color Guide" on page 41 for placement. (For general instructions, see "Stenciling" on page 13.)

⑥ Use an embossing tool to apply Moon Yellow La De Da Dots to centers of all flowers and Black La De Da Dots around the flowers on the window frame. (See "La De Da Dots" on page 16.)

⑦ Sign your project. Allow the paint to dry for several days. Apply at least three coats of varnish to protect your work. (See "Final Touches" on page 17.)

⑧ Measure the window opening. Use a marking pen, metal ruler, and glass cutter to cut a pane of glass to fit the opening. (See "Working with Glass" on page 11.) Install the glass, securing it with silicone. Allow to dry, following the manufacturer's recommendations.

⑨ Install the shelf inside the cabinet by screwing it in place. Attach the doorknob and faceplate to the window frame. Hinge the window frame to the cabinet. Attach picture hangers to the back of the cabinet.

Vintage Showcase of Flowers
Wall Art

When you practice the art of recycling, you explore every opportunity that comes along. Stopping by a remodeling project to see if there were any windows or doors to salvage, Judy met a worker who had his own personal stash of windows. He was willing to part with them for the right price, and he and Judy struck a deal. This window comes from that collection.

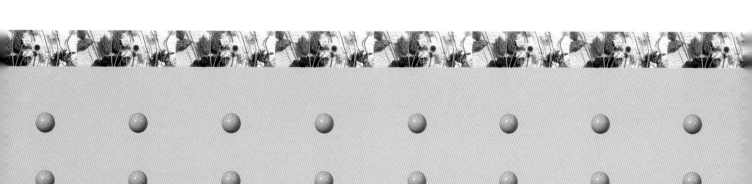

Tools and Supplies

1 four-paned window frame (ours is 27" x 36")

Glass

Assorted dried flowers

Vintage lace curtain

Foam core board

2 eye hooks

18-gauge wire

Acrylic paints (see below)

Crackle medium

Clear silicone

Spray adhesive

Foam core knife

Glass cutter

Marking pen

Packaging tape

Staple gun and staples

Straight edge metal ruler

General tools and supplies

Paints

We used the following paints to create our "Vintage Showcase of Flowers" project. (See "Suppliers" on page 94.) For a different look, substitute the colors of your choice.

DecoArt Americana acrylic paints:

Dried Basil Green (DA198)

Light Parchment (DA243)

Instructions

1 Use a hammer to break out the existing glass, if any, from your window frame. (See "Working with Glass" on page 11.)

2 Prepare the window frame for painting. (See "Surface Preparation" on page 11.)

3 Paint the window frame with Dried Basil Green. Apply as many base coats of paint as necessary to achieve smooth, opaque coverage. (See "Base Coat Painting" on page 12).

4 Apply crackle medium to the entire window frame and allow it to dry completely. Then apply a coat of Light Parchment over the crackle medium. (See "Crackle Finishes" on page 15.)

5 Sign your project. Allow the paint to dry for several days. Apply at least three coats of varnish to protect your work. (See "Final Touches" on page 17.)

6 Measure the window openings. Use a marking pen, metal ruler, and glass cutter to cut a pane of glass for each opening. (See "Working with Glass" on page 11.) Install the glass panes, securing them with silicone. Allow to dry, following the manufacturer's recommendations.

7 Use a foam core knife to cut a piece of foam core board to fit behind the glass panes and partway onto the surrounding wooden framing. Use spray adhesive to mount the lace on the foam core board. Use scissors to cut off any excess lace.

8 Lay the window frame facedown. Place a loose bouquet of dried flowers on each glass pane. Lay the foam core board on top, lace side down, sandwiching the flowers in between. Staple around the outside edge. Seal the edge with packaging tape.

9 Attach eye hooks and 18-gauge wire for hanging.

Elegance on the Wild Side
Jelly Cabinet

On a "junk run" out of town, Judy came across a Goodwill store that had its own salvage yard. This window came from a collection purchased at that store. When you find more than one window at a time, the price is good, and your wallet is fat, it doesn't hurt to purchase the lot.

Tools and Supplies

1 single-paned window frame (ours is 18" x 36")

Pine lumber (see step 2)

¼" plywood

4 wooden ball knobs, 3" in diameter

Glass

1 glass door handle

2 door hinges

Acrylic paints (see below)

Etchall Etching Creme and supplies

Faux glazing medium

Clear silicone

E6000 adhesive

Cheesecloth

3" chip brush

Embossing tool

Foam block or egg carton

Friendly woodworker or advanced woodworking tools

Glass cutter

Latex gloves

Marking pen

Measuring cup

1-quart mixing tub and stirring stick

Stencils (see below)

Straight edge metal ruler

General tools and supplies

Paints and Stencils

We used the following paints and stencils to create our "Elegance on the Wild Side" project. (See "Suppliers" on page 94.) For a different look, substitute the colors or stencils of your choice.

DecoArt Americana acrylic paints:

Light Buttermilk (DA164)—10 bottles

Camel (DA191)—3 bottles

Honey Brown (DA163)

Violet Haze (DA197)

Wisteria (DA211)

Black (DA67)

Stencilled Garden stencils:

Animal Print (TSG127)

Bella Florish Border (TSG530L)

Gingham (TSG112L) for glass etching

Stencil Color Guide

Animal Print: Wisteria, Violet Haze, Black

Bella Florish Border: Honey Brown

Base Coat Color Guide

Window frame, top of cabinet, ball knobs: Violet Haze

Cabinet (inside and outside): Light Buttermilk

Instructions

1. Use a hammer to break out the existing glass, if any, from your window frame. (See "Working with Glass" on page 11.)

2. Measure the window frame to determine the width and length of the jelly cabinet. To make the cabinet, you'll need 1 back, 2 sides, 1 top, 1 bottom, and 2 shelves, all cut from ¼" plywood; and 4 front trim, 1 bottom trim, and 2 side trim pieces, all cut from pine. Calculate the dimensions for each piece, factoring in the wood thickness, and cut the pieces with a table saw. Use a router to create decorative edges on the top and on the bottom trim pieces. Assemble the cabinet as shown. (We recruited a woodworker to make the cabinet for us.)

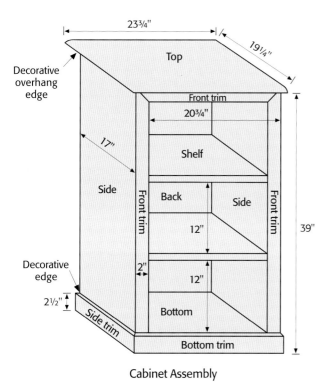

Cabinet Assembly

3. Prepare the window frame, cabinet, and wooden ball knobs for painting. (See "Surface Preparation" on page 11.)

4. Paint the wood pieces with Violet Haze and Light Buttermilk. (See "Base Coat Color Guide" on page 48). Apply as many coats as necessary to achieve smooth, opaque coverage. (See "Base Coat Painting" on page 12.) Use cotton swabs and a foam block to hold the knobs upright. Allow all the pieces to dry completely.

5. Apply a coat of varnish to the Light Buttermilk sections of the cabinet. Allow the varnish to dry completely.

6. Mask off the top of the cabinet with painter's tape. Use a measuring cup to measure and pour 16 ounces of faux glazing medium into the mixing tub. Add 4 ounces of Camel paint and mix with a stirring stick. Use a chip brush to spread the glaze mixture over a small section of the varnished area. Dab a wadded-up piece of cheesecloth over the surface to eliminate the brush strokes and stipple the glaze. Continue applying and stippling the glaze, small areas at a time, over the entire varnished surface. (See "Negative Glazed Finishes" on page 16.) Allow the glaze to dry completely. Remove the tape.

7. Stencil the designs, referring to the project photo on page 46 and the "Stencil Color Guide" on page 48 for placement. (For general instructions, see "Stenciling" on page 13.)

8. Use an embossing tool to apply Camel La De Da Dots to the outside edge of the top of cabinet and to the wooden ball knobs. (See "La De Da Dots" on page 16.)

9. Sign your project. Allow all the paint to dry for several days. Apply at least three coats of varnish to protect your work. (See "Final Touches" on page 17.)

10. Measure the window frame opening. Use a marking pen, metal ruler, and glass cutter to cut a piece of glass the same size. (See "Working with Glass" on page 11.) Install the

glass pane in the window frame, securing it around the edges with silicone. Allow the silicone to dry completely.

11. Read the Etchall Etching Creme manufacturer instructions and set up your work area. Use a spray cleaner and paper towels to thoroughly clean and dry the glass. Apply Con-Tact paper to the entire surface. Place the Gingham stencil on top of the Con-Tact paper and trace the design. Move the stencil around to mark the entire surface. Cut on the marked lines with an X-Acto knife. Peel away the paper to reveal the cutout sections. Press out any air bubbles in the remaining mask. Burnish the edges.

12. Put on goggles and latex gloves. Apply the etching cream to the glass and leave it in place the required time. (We applied the medium with a squeegee and left it on the glass for 15 minutes. We then scraped the excess medium back into its container for reuse.) Still wearing goggles and gloves, rinse the glass thoroughly under running water to remove the Con-Tact paper and any residual etching medium. Wash and dry the glass.

13. Use E6000 adhesive to attach the knob "feet" to the bottom of the cabinet.

14. Attach the door handle to the window frame. Hinge the window frame to the cabinet for a door.

Whimsy Garden
Potting Table

It's always good to let everyone know what you do for a hobby or living—you never know what may fall your way. Jennifer received a handsome collection of screen doors from the owner of the printing company that creates her catalog. Removed from the owner's home, the doors had been in storage at the printing company for some time. Nobody knew what to do with them, but they couldn't bear to throw them away. It was Jennifer's representative who thought of her. This potting table is one of the projects that resulted.

Tools and Supplies

1 screen door (ours is 36" x 80½")

2 wooden spindles (ours are 38" long)

4 decorative hooks

3 decorative knobs

Pine lumber (see step 2)

Wood screws

Acrylic paints (see below)

Gesso

Embossing tool

Friendly woodworker or advanced
   woodworking tools

Metal primer

Stencils (see below)

General tools and supplies

Paints and Stencils

We used the following paints and stencils to
create our "Whimsy Garden" project. (See
"Suppliers" on page 94.) For a different look,
substitute the colors or stencils of your choice.

DecoArt Americana acrylic paints:

Reindeer Moss Green (DA187)—8 bottles

Light Buttermilk (DA164)—2 bottles

Wisteria (DA211)

French Vanilla (DA184)

Black (DA67)

Violet Haze (DA197)

Moon Yellow (DA07)

True Ochre (DA143)

French Mauve (DA186)

Mauve (DA26)

Antique Maroon (DA160)

Soft Lilac (DA237)

Country Blue (DA41)

Summer Lilac (DA189)

Pansy Lavender (DA154)

Santa Red (DA170)

Hauser Light Green (DA131)

Plantation Pine (DA113)

Easy Blend Charcoal Grey (DEB28)

White (DA01)

Stencilled Garden stencils:

Lazy Dazy Flower (TSG733)

Swirlie Flower (TSG731)

Polka-Dot Flower (TSG732)

Stems & Leaves (TSG740)

Bumble Bee (TSG729)

Missy Miss Lady Bug (TSG735)

Checked Butterfly (TSG736)

Folk Art Tulips (TSG240)

JF Seed Co. (TSG236)

Garden Gloves (TSG213)

Garden Critters (TSG140)

Wild Posies (TSG190)

Checkerboards (TSG706)

Mouser (TSG722)

Stencil Color Guide

Lazy Dazy Flower: Wisteria, Violet Haze, Moon Yellow, True Ochre

Swirlie Flower: French Mauve, Mauve, Moon Yellow, True Ochre, Antique Maroon

Polka-Dot Flower: Soft Lilac, Country Blue, Moon Yellow, True Ochre

Stems & Leaves: Hauser Light Green, Plantation Pine

Bumble Bee: Moon Yellow, True Ochre, Black, Easy Blend Charcoal Grey

Missy Miss Lady Bug: Santa Red, Black

Checked Butterfly: Summer Lilac, Pansy Lavender, Black

Folk Art Tulips: Wisteria, Violet Haze, Hauser Light Green, Plantation Pine

JF Seed Co.: French Vanilla, White, Santa Red, Hauser Light Green, Plantation Pine, Black, Wisteria, Violet Haze, Country Blue

Garden Gloves: Light Buttermilk, Wisteria, Violet Haze

Garden Critters: Santa Red, Black

Flowers from Wild Posies: Wisteria, Violet Haze, Hauser Light Green, Plantation Pine

Checkerboards, ¾" and 1": Black

Mouser: Wisteria, Violet Haze

Base Coat Color Guide

Screen door frame, table top, and bottom shelf: Reindeer Moss Green

Screen: Light Buttermilk

Spindles (top to bottom): Wisteria, Reindeer Moss Green, French Vanilla, Black, Violet Haze

Instructions

1. Measure the screen door to determine the width of the potting table. To make the table, you'll need 1 top, 1 backsplash, 1 front apron, 2 side aprons, and 1 shelf, all cut from pine, plus 2 spindles for the table legs. Calculate the dimensions for each pine piece, factoring in the wood thickness and spindle length. Note that the top and the shelf require several pieces of pine to make up the width. Cut the pieces with a table saw. Assemble the potting bench as shown; leave off the legs and shelf, which will be added after stenciling. (We recruited a woodworker to make the bench for us.)

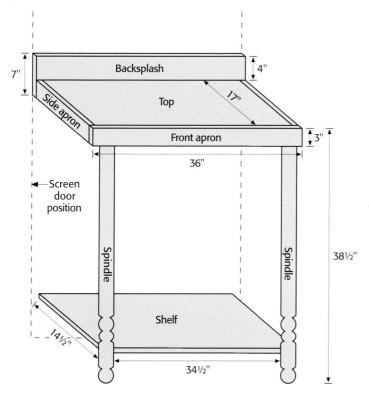

Potting Bench Assembly

2. Prepare the screen door and potting table for painting. (See "Surface Preparation" on page 11.) Also prepare the wire mesh screen with metal primer.

3. Paint the wooden pieces, referring to the color photo on page 52 and the "Base Coat Color Guide" on page 54. Apply as many coats as necessary to achieve smooth, opaque coverage. (See "Base Coat Painting" on page 12.)

4. On the spindles, use gesso to white out areas on the Black base coat that you plan to stencil. (See the section on whiting out on page 14.)

5. Stencil the designs, referring to the project photo on page 52 and the "Stencil Color Guide" on page 54 for placement. (For general instructions, see "Stenciling" on page 13.) To "overlap" the seed packets, first stencil the packet that will appear to be on top. After the paint has dried, shield the stenciled packet by placing the solid, positive (or "fallout") portion of the stencil over it. Position the second seed packet stencil on top and stencil it. Using the fallout will make this packet appear to be under the first one. Let the paint dry. To create the shadows around the packets, cover both packets with their fallouts and stencil around the edges with Easy Blend Charcoal Grey.

6. Use an embossing tool to apply Black La De Da Dots around the flowers on the seed packet, Light Buttermilk La De Da Dots to the flower centers on the gloves, Violet Haze La De Da Dots randomly on the background of the gloves, White La De Da Dots around the cherries, and Reindeer Moss Green La De Da Dots on the bottom section of the table legs. (See "La De Da Dots" on page 16.)

7. Sign your project. Allow the paint to dry for several days. Apply at least three coats of varnish to protect your work. (See "Final Touches" on page 17.)

8. Attach the knobs and hooks as shown. Attach the bench to the screen door using wood screws.

Funky Chicken
Wall Art

When you're a junker at heart, you never stop junking. Our city sponsors a couple of days each year when everyone kicks their garbage to the curb. Now, these events are a junker's heaven. The window screen for this project was saved from the trash pickup—one less piece for the pile.

Tools and Supplies

1 six-paned screen window (ours is 47" x 48")

¼" plywood

Picture hangers

Acrylic paints (see below)

Industrial strength spray adhesive

Embossing tool

Friendly woodworker or advanced woodworking tools

Stencils (see below)

General tools and supplies

Paints and Stencils

We used the following paints and stencils to create our "Funky Chicken" project. (See "Suppliers" on page 94.) For a different look, substitute the colors or stencils of your choice.

DecoArt Americana acrylic paints:

Olive Green (DA56)—6 bottles

Black (DA67)

Moon Yellow (DA07)

Santa Red (DA170)

Napa Red (DA165)

Hauser Light Green (DA131)

Plantation Pine (DA113)

White (DA01)

Violet Haze (DA197)

Wisteria (DA211)

Soft Lilac (DA237)

Country Blue (DA41)

Marigold (DA194)

Charcoal Grey (DA88)

Stencilled Garden stencils:

Zebra Print (TSG129)

Penny's Poultry Gone Wild (TSG142)

Chickie Eggs (TSG727)

Chick Chick Wire (TSG726)

Kitty Kat Daisy's (TSG191)

Girly's Flowers (TSG175)

Wild Posies (TSG190)

Squiggles & Dots (TSG178)

Stencil Color Guide

Zebra Print: Black

Penny's Poultry Gone Wild: Moon Yellow, Santa Red, Napa Red, Violet Haze, Black, Wisteria, Soft Lilac, Country Blue

Chickie Eggs: Moon Yellow, Wisteria, Violet Haze, Soft Lilac, Country Blue

Chick Chick Wire: Charcoal Grey

Flowers from Kitty Kat Daisy's: White

Girly's Flowers: Soft Lilac, Country Blue, Hauser Light Green, Plantation Pine, Santa Red, Napa Red, Moon Yellow, Marigold

Flowers from Wild Posies: Santa Red, Napa Red, Hauser Light Green, Plantation Pine, Soft Lilac, Country Blue, Moon Yellow, Marigold, Wisteria, Violet Haze

Squiggles from Squiggles & Dots: Violet Haze

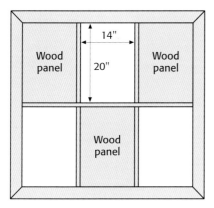

Instructions

1 Measure the window screen openings (ours are 14" x 20" each). Use a table saw to cut plywood panels to fit three of the openings. (We recruited a woodworker to do this step for us.) Use spray adhesive to attach the panels as shown.

Attach the wood panels.

2 Prepare the window screen frame and wood panels for painting. (See "Surface Preparation" on page 11.) We left the metal screen as we found it.

3 Paint the frame and panels with Olive Green. Apply as many coats as necessary to achieve smooth, opaque coverage. (See "Base Coat Painting" on page 12.)

4 Stencil the designs, referring to the project photo on page 56 and the "Stencil Color Guide" on page 58 for placement. (For general instructions, see "Stenciling" on page 13.)

5 Use an embossing tool to apply Moon Yellow and Soft Lilac La De Da Dots on the flower centers and Black La De Da Dots randomly around flowers. (See "La De Da Dots" on page 16.)

6 Sign your project. Allow the paint to dry for several days. Apply at least three coats of varnish to protect your work. (See "Final Touches" on page 17.) Attach picture hangers to the back.

Hanging Out in the Garden
Window Box

Through years of being in business, we've developed some great relationships. Just about everyone knows that we love to save lost objects. One particular owner of a screen company collects the old screens that he's replacing, and for a fair price, he delivers the load to us. We couldn't have made this planter box without him.

Tools and Supplies

1 wood-framed window screen (ours is 30" x 34")

Pine lumber (see step 1)

4 wooden ball knobs, 2¼" in diameter

Acrylic paints (see below)

E6000 adhesive

Friendly woodworker or advanced woodworking tools

Embossing tool

Stencils (see below)

General tools and supplies

Paints and Stencils

We used the following paints and stencils to create our "Hanging Out in the Garden" project. (See "Suppliers" on page 94.) For a different look, substitute the colors or stencils of your choice.

DecoArt Americana acrylic paints:

Moon Yellow (DA07)—2 bottles

Black (DA67)

Santa Red (DA170)

Wisteria (DA211)

Violet Haze (DA197)

Hauser Light Green (DA131)

Plantation Pine (DA113)

Taffy Cream (DA05)

Soft Lilac (DA237)

Country Blue (DA41)

Stencilled Garden stencils:

Girly's Gone Checked (TSG222)

Missy Miss Lady Bug (TSG735)

Girly's Flowers (TSG175)

Squiggles & Dots (TSG178)

Swirlie Flower (TSG731)

Puffy Flower (TSG734)

Little Checks (TSG707)

Petal Flower (TSG730)

Stencil Color Guide

Checks from Girly's Gone Checked: Black

Missy Miss Lady Bug: Santa Red, Black

Girly's Flowers: Wisteria, Violet Haze, Hauser Light Green, Plantation Pine

Squiggles from Squiggles & Dots: Violet Haze

Swirlie Flower: Wisteria, Violet Haze

Puffy Flower: Santa Red, Black

Little Checks, ¼": Taffy Cream, Moon Yellow

Petal Flower: Soft Lilac, Country Blue, Moon Yellow

Base Coat Color Guide

Window screen and flower box: Moon Yellow

Swirlie Flower: Wisteria

Puffy Flower and wooden ball knobs: Santa Red

Petal Flower: Soft Lilac

Instructions

1. Measure the window screen to determine the length of the flower box. To make the box, you'll need 1 front, 1 back, 2 sides, and 1 bottom, all cut from pine. Calculate the dimensions for each piece, factoring in the wood thickness, and cut the pieces with a table saw. Assemble the box as shown. (We recruited a woodworker to make the box for us.)

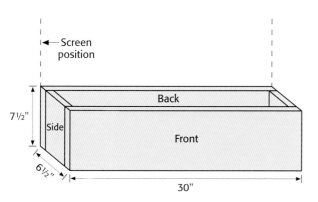

Flower Box Assembly

2. Use the stencils to trace 1 Petal, 1 Puffy, and 1 Swirlie flower onto pine. Use a jigsaw or coping saw to cut out each shape. (Our woodworker friend helped us here too.)

3. Prepare the window screen, flower box, cutout flowers, and ball knobs for painting. (See "Surface Preparation" on page 11.)

4. Paint the flower box and frame with Moon Yellow, the Swirlie flower with Wisteria, the Petal flower with Soft Lilac, and the Puffy flower and ball knobs with Santa Red. Refer to the project photo and the "Base Coat Color Guide" on page 62 for the color placement. Apply as many coats as necessary to achieve smooth, opaque coverage. (See "Base Coat Painting" on page 12.)

5. Stencil the designs, referring to the project photo on page 60 and the "Stencil Color Guide" on page 62. (For general instructions, see "Stenciling" on pages 13.)

6. Use an embossing tool to apply Black La De Da Dots randomly around Girly's flowers, Swirlie flower, and Puffy flower, and to the wooden ball knobs. Apply Country Blue La De Da Dots to Petal flower. (See "La De Da Dots" on page 16.)

7. Sign your project. Allow the paint to dry for several days. Apply at least three coats of varnish to all parts to protect your work. (See "Final Touches" on page 17.)

8. Use E6000 adhesive to attach the knobs to the bottom of the flower box. Glue the flower cutouts to the window screen. Screw the flower box to screen.

Tyler's
Game @ 4:30

Dinner @ 7pm

Milk
Bread
eggs
Butter

Painted Whimsies: Decorative Accents
For Your Home and Garden
Join authors Jennifer Ferguson and Judith Skinner
For a Demonstration and Book Signing
Tuesday, May 20th at 7:00 pm

Jazz Up Your Mailbox. Perk Up Those Plant Pots!
Learn How To Breath New Life
Into The Most Ordinary Items Tonight!

Painted Chairs by Jennifer Ferguson
And Judith Skinner Will Also
Be Available.

Leave a Message
Blackboard and Corkboard

When we attended a painting convention in San Diego, we set aside a little time for some junking. Who can resist exploring new territory? That's when we came across the Architectural Salvage Shop. This set of cabinet doors was purchased there, along with several other special finds.

Tools and Supplies

2 cabinet doors (ours are 18" x 12" each)

2 decorative knobs

Pine lumber (see step 1)

Corkboard

Acrylic paints (see below)

Chalkboard paint

Embossing tool

E6000 adhesive

Friendly woodworker or advanced woodworking tools

Picture hanger

Stencils (see below)

Wood screws

General tools and supplies

Paints and Stencils

We used the following paints and stencils to create our "Leave a Message" project. (See "Suppliers" on page 94.) For a different look, substitute the colors or stencils of your choice.

DecoArt Americana acrylic paints:

Santa Red (DA170)—3 bottles

Reindeer Moss Green (DA187)

Celery Green (DA208)

Violet Haze (DA197)

Moon Yellow (DA07)

Marigold (DA194)

Black (DA67)

Hauser Light Green (DA131)

Plantation Pine (DA113)

Stencilled Garden stencils:

Java Time (TSG198)

Wild Posies (TSG190)

Stencil Color Guide

Java Time: Reindeer Moss Green, Celery Green, Violet Haze, Moon Yellow, Marigold, Black

Flowers from Wild Posies: Moon Yellow, Marigold, Hauser Light Green, Plantation Pine

Instructions

1. Choose a strip of pine that is 1" deeper than one cabinet door. Cut the pine strip the same length as the door's lower edge. Align the strip along the lower edge, flush at the back and protruding 1" at the front, to form a ledge for chalk. Attach the ledge to the door using wood screws and glue. (We recruited a woodworker to do this step for us.)

2. Prepare both doors for painting. (See "Surface Preparation" on page 11.)

3. Paint both doors with Santa Red, going in about 1" on each inset panel. Apply as many coats as necessary to achieve smooth, opaque coverage. (See "Base Coat Painting" on page 12.)

4. Apply chalkboard paint to the inset panel of the door with the ledge. Follow the paint manufacturer's instructions.

5. Stencil the designs, referring to the project photo on page 64 and the "Stencil Color Guide" above. (For general instructions, see "Stenciling" on page 13.)

6. Use an embossing tool to apply Violet Haze La De Da Dots to the flower centers and Black La De Da Dots randomly around the flowers. (See "La De Da Dots" on page 16.)

7. Sign your project. Allow the paint to dry for several days. Apply at least three coats of varnish to protect your work. (See "Final Touches" on page 17.)

8. Measure the remaining door inset. Use scissors to cut a matching piece from corkboard. Glue the corkboard in place with the E6000 adhesive. Attach a decorative knob to each door. Attach picture hangers to the back.

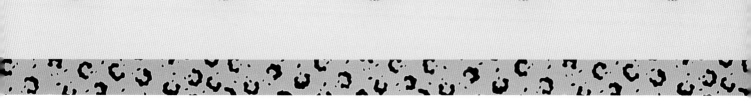

The Call of the Jungle
Hall Mirror

Everyone makes mistakes, even our local cabinetmaker. This cabinet door was a mistake for someone (wrong dimensions or something like that) but an opportunity for Jennifer. She was lucky enough to receive a large pile of mistakes in various sizes for a very fair price.

Tools and Supplies

1 cabinet door (ours is 16" x 27")

5 decorative knobs

Mirror

Picture hangers

Acrylic paints (see below)

Gunther glue (we generally use Ultra/bond)

Embossing tool

Glass cutter

Marking pen

Stencils (see below)

Straight edge metal ruler

Watercolor pencil (optional)

General tools and supplies

Paints and Stencils

We used the following paints and stencils to create our "The Call of the Jungle" project. (See "Suppliers" on page 94.) For a different look, substitute the colors or stencils of your choice.

DecoArt Americana acrylic paints:

Reindeer Moss Green (DA187)—2 bottles

Deep Periwinkle (DA212)

Black (DA67)

Wisteria (DA211)

Celery Green (DA208)

Plantation Pine (DA113)

Stencilled Garden stencils:

Zebra Print (TSG129)

Giraffe Print (TSG230)

Wild Animal Print (DA128)

Animal Print (DA127)

Ashley's Tea Party (DA183)

Stencil Color Guide

Zebra Print: Black

Giraffe Print: Black

Wild Animal Print: Black

Animal Print: Celery Green, Black

Flowers from Ashley's Tea Party: Wisteria, Celery Green, Plantation Pine

Instructions

1. Prepare the cabinet door for painting. (See "Surface Preparation" on page 11.)

2. Paint the cabinet door with Reindeer Moss Green. Go about 1" in on the panel that will hold the mirror. Apply as many coats of paint as necessary to achieve smooth, opaque coverage. (See "Base Coat Painting" on page 12.) Allow the paint to dry thoroughly.

3. Use painter's tape to diagonally mask off all four corners of the door. Burnish the tape edges. Paint the masked-off corners with Deep Periwinkle, as shown in the project photos on pages 68 and 70. Allow the paint to dry before removing the tape. (These sections can also be painted freehand, if you prefer. Use a watercolor pencil to draw faint guidelines.)

4. Stencil the designs, referring to the project photo on page 68 and the "Stencil Color Guide" above for placement. (For general instructions, see "Stenciling" on page 13.)

5. Use an embossing tool to apply Black La De Da Dots randomly around the flowers and along the diagonal lines at the four corners. (See "La De Da Dots" on page 16.)

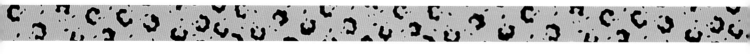

6 Sign your project. Allow the paint to dry for several days. Apply at least three coats of varnish to protect your work. (See "Final Touches" on page 17.)

7 Measure the inner panel of the door. If the opening is a standard size, you may be able to purchase a beveled edge mirror to fit it, as we did. Otherwise, use a marking pen, metal ruler, and glass cutter to cut a piece of mirror the same size. (See "Working with Glass" on page 11.) Install the mirror panel with Gunther glue, following the manufacturer's instructions and recommended cure time.

8 Attach picture hangers to the back of the door. Attach decorative knobs to the front bottom edge.

Harlequin Box
Medicine Cabinet

Our friend who deals in salvage calls us on a regular basis. (To be honest, he drops by in his pickup and we go "shopping" out of the back of his truck in Jennifer's parking lot.) This cabinet is a piece that Jennifer purchased from him many years ago. Her updated version features ceiling tin on the outside door panel and a mirror on the inside.

Tools and Supplies

1 bathroom medicine cabinet (ours is 24" x 28½")

Ceiling tin

1 large finial

2 small finials

4 wooden ball knobs, 2¼" diameter

Mirror (for back of cabinet door)

Heavy-duty picture hangers

Acrylic paints (see below)

Metal primer

Gunther glue

E6000 adhesive

Embossing tool

Glass cutter

Marking pen

Stencils (see below)

Straight edge metal ruler

White tapered candle

General tools and supplies

Paints and Stencils

We used the following paints and stencils to create our "Harlequin Box" project. (See "Suppliers" on page 94.) For a different look, substitute the colors or stencils of your choice.

DecoArt Americana acrylic paints:

Yellow Ochre (DA08)—4 bottles

Napa Red (DA165)—3 bottles

Black (DA67)

White (DA01)

Country Blue (DA41)

Celery Green (DA208)

Plantation Pine (DA113)

Stencilled Garden stencils:

Connect the Dots (TSG246S)

Wild Posies (TSG190)

Stencil Color Guide

Connect the Dots: Napa Red, Black

Flowers from Wild Posies: Yellow Ochre, Celery Green, Plantation Pine

Base Coat Color Guide

Cabinet base and door frame: Yellow Ochre

Cabinet door ball knobs: Napa Red

Finials (top to bottom): Napa Red, Yellow Ochre, Black

Instructions

1. Prepare the cabinet for painting. (See "Surface Preparation" on page 11.)

2. Paint the cabinet, finials, and ball knobs with Yellow Ochre, Napa Red, and Black. Refer to the project photo on page 72 and the "Base Coat Color Guide" above for the color placement. Apply as many coats as necessary to achieve smooth, opaque coverage. (See "Base Coat Painting" on page 12.)

3. Apply metal primer to the ceiling tin. Allow to dry completely. Randomly paint Yellow Ochre and Country Blue on the primed tin. Allow to dry completely.

4. To create a peeled paint look, rub the candle on the painted tin in selected areas, pressing hard so that the wax transfers and sticks. Apply two coats of White paint to the entire piece, allowing each coat to dry completely. To expose the Yellow Ochre and Country Blue base coat, sand the tin with 100-grit sandpaper, concentrating on the areas where you applied the wax. Expose as much of the base coat colors as you desire. The wax prevents the top coats of paint from sticking and makes the sanding process easier. Use a tack cloth to wipe off the dust.

5 Stencil the designs, referring to the project photo on page 72 and the "Stencil Color Guide" on page 73 for placement. (For general instructions, see "Stenciling" on page 13.)

6 Use an embossing tool to apply Napa Red La De Da Dots to the flower centers, White La De Da Dots randomly around the flowers, and Yellow Ochre La De Da Dots to the wooden ball knobs. (See "La De Da Dots" on page 16.)

7 Sign your project. Allow the paint to dry for several days. Apply at least three coats of varnish to protect your work. (See "Final Touches" on page 17.)

8 Measure the panel inset on the reverse side of the cabinet door. Use a marking pen, metal ruler, and glass cutter to cut a piece of mirror the same size. (See "Working with Glass" on page 11.) Install the mirror with Gunther glue, following the manufacturer's instructions and recommended cure time.

9 Using old scissors, trim the ceiling tin to fit on the outside panel of the cabinet door. Use E6000 adhesive to glue on the tin, knobs, and finials.

10 Attach the picture hangers to the back.

Girly's Privacy
Folding Screen

Okay, so we're not junking all the time—or are we? Everybody needs to stay in shape, and walking the dog is a good form of exercise. Look what Belle sniffed out in the trash. When Judy saw these louvered shutter panels, she grabbed 'em. Like we keep saying over and over again—keep your eyes open wherever you go, even if it's just to walk the dog around the block.

Tools and Supplies

2 shutter-style closet doors (our panels are 15" x 78" each)

Decorative door handle

Acrylic paints (see below)

Stencils (see below)

General tools and supplies

Paints and Stencils

We used the following paints and stencils to create our "Girly's Privacy" project. (See "Suppliers" on page 94.) For a different look, substitute the colors or stencils of your choice.

DecoArt Americana acrylic paints:

Light Buttermilk (DA164)—8 bottles

Black (DA67)

Hauser Light Green (DA131)

Plantation Pine (DA113)

Khaki Tan (DA173)

Burnt Umber (DA64)

Pansy Lavender (DA154)

Summer Lilac (DA189)

Reindeer Moss Green (DA187)

Marigold (DA194)

Easy Blend Charcoal Grey (DEB28)

Soft Lilac (DA237)

Country Blue (DA41)

Stencilled Garden stencils:

Girly's Topiary (TSG244)

Bee Happy (TSG177)

Butterflies (TSG701)

Checkerboards (TSG706)

Stencil Color Guide

Girly's Topiary: Hauser Light Green, Plantation Pine, Reindeer Moss Green, Summer Lilac, Pansy Lavender, Khaki Tan, Burnt Umber, Black

Bees from Bee Happy: Marigold, Black, Easy Blend Charcoal Grey

Butterflies: Soft Lilac, Country Blue, Black

Checkerboards, 1": Black

Instructions

1. Prepare the shutter doors for painting. (See "Surface Preparation" on page 11.)

2. Paint the shutters with Light Buttermilk. Apply as many base coats of paint as necessary to achieve smooth, opaque coverage. (See "Base Coat Painting" on page 12.) Allow the paint to dry completely.

3. For a worn/aged look, sand the entire piece with medium to fine sandpaper, exposing old paint or raw wood wherever you desire.

4. Stencil the designs, referring to the project photo on page 76 and the "Stencil Color Guide" above for placement. (For detailed instructions, see "Stenciling" on page 13.)

5. Sign your project. Allow all the paint to dry for several days. Apply at least three coats of varnish to protect your work. (See "Final Touches" on page 17.)

6. Attach a decorative door handle.

You've Got Mail
Display Stand

W hen your own stockpile of recyclables is running low, check out a friend's treasure trove. Desperately in need of a pair of shutters, Jennifer went shopping at Judy's warehouse. She found shutters and so much more.

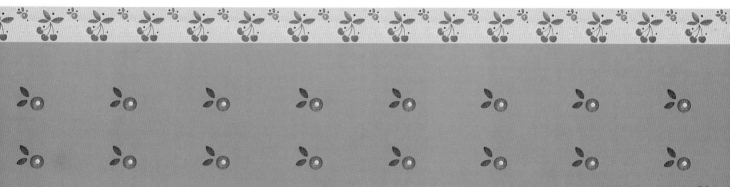

Tools and Supplies

2 shutter-style closet doors (ours are 11" x 78" each)

5 decorative knobs

¼" plywood

Wood screws

Acrylic paints (see below)

Embossing tool

Friendly woodworker or advanced woodworking tools

Stencils (see below)

General tools and supplies

Paints and Stencils

We used the following paints and stencils to create our "You've Got Mail" project. (See "Suppliers" on page 94.) For a different look, substitute the colors or stencils of your choice.

DecoArt Americana acrylic paints:

French Vanilla (DA184)—8 bottles

Santa Red (DA170)

Napa Red (DA165)

Hauser Light Green (DA131)

Plantation Pine (DA113)

Soft Lilac (DA237)

Country Blue (DA41)

Black (DA67)

Stencilled Garden stencils:

Summertime (TSG176)

Ashley's Tea Party (TSG183)

Cherries Jubillee (TSG184)

Stencil Color Guide

Ladybugs from Summertime:
Santa Red, Black

Flowers from Ashley's Tea Party:
Soft Lilac, Country Blue, Hauser Light Green, Plantation Pine

Cherries Jubillee: Santa Red, Napa Red, Hauser Light Green, Plantation Pine

Instructions

1. Measure the louvered section of each shutter. Use a table saw to cut a piece of plywood 1" wider and 1" longer to back each louvered section. (We recruited a woodworker to do this step for us.)

2. Prepare the shutter doors and plywood panels for painting. (See "Surface Preparation" on page 11.)

3. Paint the shutters and plywood panels with French Vanilla. Apply as many base coats of paint as necessary to achieve smooth, opaque coverage. (See "Base Coat Painting" on page 12.) Allow the paint to dry completely.

4. Stencil the designs, referring to the project photo on page 78 and the "Stencil Color Guide" above for placement. (For detailed instructions, see "Stenciling" on page 13.)

5. Use an embossing tool to apply French Vanilla La De Da Dots to the flower centers and Black La De Da Dots randomly around the flowers and cherries and in groups of three around the outside edge of shutters. (See "La De Da Dots" on page 16.)

6. Sign your project. Allow the paint to dry for several days. Apply at least three coats of varnish to protect your work. (See "Final Touches" on page 17.)

7. Screw the plywood panels to the backs of the shutters. Attach the decorative knobs to the front. Slip your letters and postcards between the slats.

Hidden Treasures
Rolling Under-Bed Storage

It never hurts to see what your friends and family might be discarding after a move. While visiting a friend who had recently relocated, Jennifer's keen eye noticed these old cabinet drawers. She figured that someday we would turn them into something special. This practical storage unit was just the ticket.

Tools and Supplies

2 cabinet drawers (ours are 12" x 23" x 5" each)

2 decorative drawer handles

4 casters

Wood screws

Acrylic paints (see below)

Embossing tool

Stencils (see below)

General tools and supplies

Paints and Stencils

We used the following paints and stencils to create our "Hidden Treasures" project. (See "Suppliers" on page 94.) For a different look, substitute the colors or stencils of your choice.

DecoArt Americana acrylic paints:

Light Buttermilk (DA164)—6 bottles

Buttermilk (DA03)

Wisteria (DA211)

Violet Haze (DA197)

Black (DA67)

Olive Green (DA56)

Hauser Light Green (DA131)

Evergreen (DA82)

Stencilled Garden stencils:

Curvy Checks (TSG715S)

Girly's Gone Checked (TSG222)

Little Checks (TSG707)

Stencil Color Guide

Curvy Checks: Black, Olive Green

Girly's Gone Checked: Wisteria, Violet Haze, Hauser Light Green, Evergreen

Little Checks, ¼": Black

Instructions

1. Place the two drawers side by side. Use an electric drill to drill starter holes and then screw the drawers together to create one unit.

2. Prepare the drawers for painting. (See "Surface Preparation" on page 11.)

3. Paint the drawers with Light Buttermilk. Apply as many base coats of paint as necessary to achieve smooth, opaque coverage. (See "Base Coat Painting" on page 12.) Allow the paint to dry completely.

4. Pour a small amount of Buttermilk paint onto your palette. Dilute the paint with water. Dip a ¾"-wide artist's brush into the paint. Working freehand, paint vertical stripes approximately 1" apart on the outer side walls. Make the stripes soft and whimsical in appearance rather than precise. A new brush with bristles that haven't yet flared will give the best results.

5. Stencil the designs, referring to the project photo on page 82 and the "Stencil Color Guide" at left for placement. (For detailed instructions, see "Stenciling" on page 13.)

6. Use an embossing tool to apply Black La De Da Dots to the top edge of the drawers and randomly around the flowers. (See "La De Da Dots" on page 16.)

7. Sign your project. Allow the paint to dry for several days. Apply at least three coats of varnish to protect your work. (See "Final Touches" on page 17.)

8. Attach casters to the bottom four corners, drilling starter holes as needed. Attach decorative drawer handles to the front.

Cherry Delight
Serving Tray

We don't only talk about recycling, we try to practice what we preach. When Jennifer's own kitchen remodeling got underway, this cabinet drawer—original to the house—was no longer needed. Now it lives on as a serving tray—and Jennifer got it for a very good price!

Tools and Supplies

1 cabinet drawer (ours is 16" x 22½" x 3½")

4 finials

2 decorative drawer handles

Acrylic paints (see below)

E6000 adhesive

Stencil (see below)

General tools and supplies

Embossing tool

Paints and Stencil

We used the following paints and stencil to create our "Cherry Delight" project. (See "Suppliers" on page 94.) For a different look, substitute the colors or stencil(s) of your choice.

DecoArt Americana acrylic paints:

Light Buttermilk (DA164)—4 bottles

Country Blue (DA41)

Santa Red (DA170)

Napa Red (DA165)

Moon Yellow (DA07)

Marigold (DA194)

Hauser Light Green (DA131)

Plantation Pine (DA113)

Stencilled Garden stencil:

Fresh Cherries (TSG223)

Stencil Color Guide

Fresh Cherries: Santa Red, Napa Red, Moon Yellow, Marigold, Hauser Light Green, Plantation Pine

Instructions

1. Prepare the cabinet drawer for painting. (See "Surface Preparation" on page 11.)

2. Paint the drawer with Light Buttermilk. Paint the finials Country Blue, Santa Red, and Moon Yellow. Apply as many base coats of paint as necessary to achieve smooth, opaque coverage. (See "Base Coat Painting" on page 12.) Allow the paint to dry completely.

3. Stencil the designs, referring to the project photo on page 84 and the "Stencil Color Guide" at left for placement. (For detailed instructions, see "Stenciling" on page 13.) Allow to dry completely.

4. Pour some Country Blue paint onto your palette. Mix in an equal amount of extender to make a wash. Using a stencil brush, apply the wash in a back-and-forth stroke around the stenciled images. Allow a little of the base coat color to show around each stenciled design.

5. Use an embossing tool to apply Country Blue La De Da Dots to the ball portion of the finials. (See "La De Da Dots" on page 16.)

6. Sign your project. Allow the paint to dry for several days. Apply at least three coats of varnish to protect your work. (See "Final Touches" on page 17.)

7. Use E6000 adhesive to attach the finials, upside down, to the bottom of the drawer. Attach the handles to opposite ends of the drawer.

Beaded Fringe
Accent Table

While Judy was visiting her granddaughter Samantha, they went to a Habitat for Humanity store, which sells home salvage items only. Judy and Samantha did a little shopping together and found the parts for this table. Samantha is turning into quite the little shopper.

Tools and Supplies

1 cabinet drawer (ours is 16" x 14" x 12½")

1 wooden newel post (ours is 37" tall)

Pine lumber (see step 1)

Glass

Beaded fringe

Acrylic paints (see below)

Gesso

Embossing tool

Friendly woodworker or advanced woodworking tools

Glass cutter

Marking pen

Staple gun and staples

Stencils (see below)

Straight edge metal ruler

General tools and supplies

Paints and Stencils

We used the following paints and stencils to create our "Beaded Fringe" project. (See "Suppliers" on page 94.) For a different look, substitute the colors or stencils of your choice.

DecoArt Americana acrylic paints:

   Soft Lilac (DA237)—4 bottles

   French Mauve (DA186)

   Raspberry (DA28)

   Olive Green (DA56)

   Khaki Tan (DA173)

   Asphaltum (DA180)

   Violet Haze (DA197)

   Wisteria (DA211)

   Black (DA67)

   Hauser Light Green (DA131)

   Plantation Pine (DA113)

   Moon Yellow (DA07)

   Marigold (DA194)

   Country Blue (DA41)

   Easy Blend Charcoal Grey (DEB28)

   Santa Red (DA170)

   White (DA01)

   French Vanilla (DA184)

   Yellow Ochre (DA08)

   Antique Gold (DA09)

   Cashmere Beige (DA91)

Stencilled Garden stencils:

   Wild Animal Print (TSG128)

   Whimsey Topiary's (TSG243)

   Miss JenJen (TSG737)

   Girly's Flowers (TSG175)

   Girly's Gone Checked (TSG222)

Stencil Color Guide

Wild Animal Print: Country Blue

Whimsey Topiary's: French Mauve, Raspberry, Khaki Tan, Asphaltum, Violet Haze, Black, Hauser Light Green, Plantation Pine, Moon Yellow, Marigold, Easy Blend Charcoal Grey

Miss JenJen: Santa Red, White, Black, Hauser Light Green, French Vanilla, Yellow Ochre, Antique Gold, Cashmere Beige, Violet Haze

Girly's Flowers: Wisteria, Violet Haze, Hauser Light Green, Plantation Pine

Girly's Gone Checked: French Mauve, Raspberry, Hauser Light Green, Plantation Pine

Base Coat Color Guide

Drawer, pedestal base: Soft Lilac

Pedestal (top to bottom): Black, Moon Yellow, Olive Green, Violet Haze, French Mauve, Black

Instructions

1 Use a jigsaw or coping saw to cut four pieces of pine for the pedestal, as shown in the diagram. The dimensions can be adjusted to suit your cabinet and newel post. Using wood screws, attach the pine pieces to the square end of the newel post, as shown in the project photo on page 88, to form the pedestal base. (We recruited a woodworker to complete this step for us.)

2½"

Leg
Cut 4 from pine.

11"

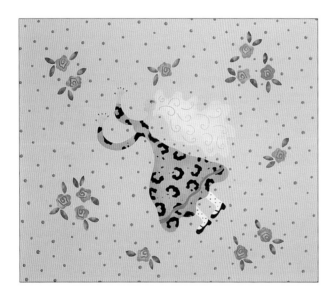

2 Prepare the cabinet drawer and pedestal for painting. (See "Surface Preparation" on page 11.)

3 Paint the drawer and pedestal with six colors, as described in the "Base Coat Color Guide" on page 89. Apply as many coats of paint as necessary to achieve smooth, opaque coverage. (See "Base Coat Painting" on page 12.)

4 Use gesso to white out areas on the Black base coat that you plan to stencil. (See the section on whiting out on page 14.)

5 Stencil the designs, referring to the project photo on page 88 and the "Stencil Color Guide" on page 89 for placement. Be sure to stencil the underside of the drawer—it will become the table top when the drawer is turned upside down. (For general instructions, see "Stenciling" on page 13.)

6 Use an embossing tool to randomly apply Country Blue La De Da Dots to the pedestal pieces and table top. Apply Black La De Da Dots on the Olive Green section of the pedestal. Apply White La De Da Dots around the flowers on the Black section of the pedestal. (See "La De Da Dots" on page 16.)

7 Sign your project. Allow the paint to dry for several days. Apply at least three coats of varnish to protect your work. (See "Final Touches" on page 17.)

8 Stand the pedestal upright. Turn the drawer upside down on top of the pedestal. Attach it securely with wood screws. Staple beaded fringe around the inside bottom edge.

9 Measure the top surface. Use a marking pen, metal ruler, and glass cutter to cut a piece of glass the same size. (See "Working with Glass" on page 11.) Set the glass on top of the table to protect your artwork.

Hooked on Diamonds
Wall-Mounted Coat Rack

Another piece recycled from Jennifer's kitchen remodeling, this cabinet drawer originally held pots and pans. She had to remove the drawer before the kitchen came down, so the pots and pans were homeless for a while.

Tools and Supplies

1 cabinet drawer front (ours is 10¼" x 32½")

5 decorative knobs

Heavy-duty picture hangers

Acrylic paints (see below)

Embossing tool

Stencils (see below)

General tools and supplies

Paints and Stencils

We used the following paints and stencils to create our "Hooked on Diamonds" project. (See "Suppliers" on page 94.) For a different look, substitute the colors or stencils of your choice.

DecoArt Americana acrylic paints:

Eggshell (DA153)

Black (DA67)

Shale Green (DA152)

Celery Green (DA208)

Plantation Pine (DA113)

Wisteria (DA211)

Violet Haze (DA197)

Light Buttermilk (DA164)

Stencilled Garden stencils:

Harliquen (TSG232)

Wild Posies (TSG190)

Stencil Color Guide

Harliquen: Shale Green, Violet Haze

Flowers from Wild Posies: Wisteria, Violet Haze, Celery Green, Plantation Pine

Instructions

1. Prepare the drawer front for painting. (See "Surface Preparation" on page 11.)

2. Paint the raised panel of the drawer front with Eggshell and the outside trim with Black. Apply as many coats of paint as necessary to achieve smooth, opaque coverage. (See "Base Coat Painting" on page 12.)

3. Stencil the designs, referring to the project photo on page 92 for placement and the "Stencil Color Guide" at left for colors. (For general instructions, see "Stenciling" on page 13.)

4. Use an embossing tool to apply Light Buttermilk La De Da Dots to the flower centers and Black La De Da Dots around the flowers. (See "La De Da Dots" on page 16.)

5. Sign your project. Allow the paint to dry for several days. Apply at least three coats of varnish to protect your work. (See "Final Touches" on page 17.)

6. Attach the decorative hardware to the front. Attach the picture hangers to the back.

Suppliers

Many of the tools and supplies described in this book are available at home improvement centers, hardware stores, craft stores, artist's supply stores, and specialty stencil stores. For more information on purchasing selected items, contact the following companies or visit the company web site:

Royal & Langnickel Brush Manufacturing, Inc.
6707 Broadway Avenue
Merrillville, IN 46410
800-247-2211

Artist's brushes

The Stencilled Garden
6029 North Palm Avenue
Fresno, CA 93704
559-449-1764
www.stencilledgarden.com

Stencils, brushes, paints, faux-painting tools and supplies, AC's Acrylic Craft Paint Remover, brush cleaner/conditioner, brush scrubbers, decorative accessories

DecoArt
www.decoart.com

Acrylic paints, glazes, gel stains

B & B Etching Products, Inc.
19721 North 98th Avenue
Peoria, AZ 85382
888-382-4255
www.etchall.com

Etchall Etching Creme

About the Authors

An artist, designer, and teacher of the arts of stenciling and faux finishing, Jennifer Ferguson has been painting for the past fourteen years. Through her company, The Stencilled Garden, she designs stencils and pattern packets, teaches stenciling and faux finishing, and sells stenciling and faux finishing supplies. Jennifer has appeared many times on *The Carol Duvall Show*, and she has also appeared on *Aleene's Creative Living* and *Kitty Bartholomew: You're Home*, where she shared some of her ideas and projects with viewers. When she isn't attending trade shows, traveling, teaching, or painting projects, Jennifer enjoys spending time with her family and working on their home.

An artist, "house stripper," and dedicated recycler, Judy Skinner has been creating art for more than twenty years. Through her own company, Collectiques by JuBee, she recycles old windows, doors, drawers, and any other house parts she can find, transforming them into works of art. She sells her unique creations at art shows throughout California and Nevada. When she isn't attending art shows or finding houses to strip, Judy enjoys spending time with her family and finishing projects for their home.

Jennifer and Judy met at Jennifer's shop back in 1996, and since then, they've developed a wonderful friendship. Their mutual love for the arts of recycling, painting, stenciling, and faux finishing gives them much to share. They travel all over the United States to attend conventions and in the process have indulged in "junking" trips that have led them to many treasures.

Trashformations is the third book for this duo, following *Painted Chairs* and *Painted Whimsies*, and they're planning many more. Jennifer and Judy both hope that you will enjoy this book and that you will fall in love with painting.

new and bestselling titles from

Martingale & COMPANY

America's Best-Loved Craft & Hobby Books®

That Patchwork Place®

America's Best-Loved Quilt Books®

NEW RELEASES
20 Decorated Baskets
Asian Elegance
Batiks and Beyond
Classic Knitted Vests
Clever Quilts Encore
Crocheted Socks!
Four Seasons of Quilts
Happy Endings
Judy Murrah's Jacket Jackpot
Knits for Children and Their Teddies
Loving Stitches
Meadowbrook Quilts
Once More around the Block
Pairing Up
Patchwork Memories
Pretty and Posh
Professional Machine Quilting
Purely Primitive
Shadow Appliqué
Snowflake Follies
Style at Large
Trashformations
World of Quilts, A

APPLIQUÉ
Appliquilt in the Cabin
Artful Album Quilts
Blossoms in Winter
Color-Blend Appliqué
Garden Party
Sunbonnet Sue All through the Year

HOLIDAY QUILTS & CRAFTS
Christmas Cats and Dogs
Christmas Delights
Creepy Crafty Halloween
Handcrafted Christmas, A
Hocus Pocus!
Make Room for Christmas Quilts
Snowman's Family Album Quilt, A
Welcome to the North Pole

LEARNING TO QUILT
101 Fabulous Rotary-Cut Quilts
Casual Quilter, The
Fat Quarter Quilts
More Fat Quarter Quilts
Quick Watercolor Quilts
Quilts from Aunt Amy
Simple Joys of Quilting, The
Your First Quilt Book (or it should be!)

PAPER PIECING
40 Bright and Bold Paper-Pieced Blocks
50 Fabulous Paper-Pieced Stars
Down in the Valley
Easy Machine Paper Piecing
For the Birds
It's Raining Cats and Dogs
Papers for Foundation Piecing
Quilter's Ark, A
Show Me How to Paper Piece
Traditional Quilts to Paper Piece

QUILTS FOR BABIES & CHILDREN
Easy Paper-Pieced Baby Quilts
Even More Quilts for Baby
More Quilts for Baby
Play Quilts
Quilts for Baby
Sweet and Simple Baby Quilts

ROTARY CUTTING/SPEED PIECING
101 Fabulous Rotary-Cut Quilts
365 Quilt Blocks a Year Perpetual Calendar
1000 Great Quilt Blocks
Around the Block Again
Around the Block with Judy Hopkins
Cutting Corners
Log Cabin Fever
Pairing Up
Strips and Strings
Triangle-Free Quilts
Triangle Tricks

SCRAP QUILTS
Nickel Quilts
Rich Traditions
Scrap Frenzy
Spectacular Scraps
Successful Scrap Quilts

TOPICS IN QUILTMAKING
Americana Quilts
Bed and Breakfast Quilts
Bright Quilts from Down Under
Creative Machine Stitching
Everyday Embellishments
Fabulous Quilts from Favorite Patterns
Folk Art Friends
Handprint Quilts
Just Can't Cut It!
Quilter's Home: Winter, The
Split-Diamond Dazzlers
Time to Quilt

CRAFTS
300 Papermaking Recipes
ABCs of Making Teddy Bears, The
Blissful Bath, The
Creating with Paint
Handcrafted Frames
Handcrafted Garden Accents
Painted Whimsies
Pretty and Posh
Sassy Cats
Stamp in Color

KNITTING & CROCHET
365 Knitting Stitches a Year
    Perpetual Calendar
Basically Brilliant Knits
Crochet for Tots
Crocheted Aran Sweaters
Knitted Sweaters for Every Season
Knitted Throws and More
Knitter's Template, A
Knitting with Novelty Yarns
More Paintbox Knits
Simply Beautiful Sweaters for Men
Today's Crochet
Too Cute! Cotton Knits for Toddlers
Treasury of Rowan Knits, A
Ultimate Knitter's Guide, The

Our books are available at bookstores and your favorite craft, fabric, and yarn retailers. If you don't see the title you're looking for, visit us at **www.martingale-pub.com** or contact us at:

1-800-426-3126

International: 1-425-483-3313 • Fax: 1-425-486-7596 • Email: info@martingale-pub.com

For more information and a full list of our titles, visit our Web site.